The ZOO That Never Was

ALSO BY R. D. LAWRENCE

Wildlife in Canada (1966)

The Place in the Forest (1967)

Where the Water Lilies Grow (1968)

The Poison Makers (1969)

Cry Wild (1970)

Maple Syrup (1971)

Wildlife in North America: Mammals (1974)

Wildlife in North America: Birds (1974)

Paddy (1977)

The North Runner (1979)

Secret Go the Wolves (1980)

Holt, Rinehart and Winston  New York

The ZOO That Never Was

R·D·Lawrence

Published by Holt, Rinehart and Winston, 383 Madison Avenue, New York, New York 10017.

Published simultaneously in Canada by Holt, Rinehart and Winston of Canada, Limited.

Library of Congress Cataloging in Publication Data
Lawrence, R. D. 1921-
The zoo that never was.
1. Wildlife rescue—Ontario. 2. Lawrence, R. D. 1921-
3. Zoologists—Ontario—Biography.
I. Title.
QL83.2.L38 639.9′092′4 80-18956
ISBN 0-03-056811-0

First Edition

Designer: Susan Mitchell
Title page art: Bill Elliott
Printed in the United States of America
10 9 8 7 6 5 4 3 2 1

For my son, Simon, and my daughter, Alison, who were briefly present at the start of my life in the wilderness and who are now sharing the result of my quest; and for my nephew, Jonathan Mitchell, whose dedication to the cause of nature should lead him along many fascinating trails.

The ZOO That Never Was

Foreword

My late wife, Joan, and I, though dedicated to conservation and to the study of natural life, did not expressly set out to become "keepers" of a wild menagerie at North Star Farm, our 350-acre property in Ontario. Indeed, the task of caring for a wide variety of orphaned or injured animals was rather thrust upon us by default in the spring of 1965, though the failure was not ours, but rather that of the authorities involved in the management of wildlife, of zoos, and of the Society for the Prevention of Cruelty to Animals. Each of these bodies expected the others to look after the problem of those wild animals that were found in need of care, in most cases because of human interference. The people who managed the wildlife of the Province of Ontario felt, not unreasonably, that they were not in the business of zoo keeping; the zoo people, who were overstocked on the so-called common animals, felt that either the wildlife people or the SPCA should look after the problem, while officials of the SPCA held the view that *their* major thrust should be directed toward caring for domestic strays, explaining that they did not have the facilities, nor the finances, to take on wild animals. The arguments expounded by each group were reasonable, but the fact remained that those wild animals unfortunate enough to come into conflict with man had nobody to champion them.

A year before the first wild strays came into our care, Joan and I had decided to marry, but because neither of us wanted to settle down to a life of urban domesticity in Winnipeg, Manitoba, where we were living at the time, we migrated to Ontario, where we hoped to be able to buy some wilderness land and to continue our work in the area of ecology. Before that, I had been something of a wilderness wanderer, accompanied by my wolf-dog Yukon,* but after his departure and following my commitment to Joan, the cares of domesticity and the urgent need to earn money caused me to buckle down under the yoke of journalism.

It was not long after we had come to live in Ontario that we found a piece of property one hundred miles north of the city of Toronto that stimulated me to write a couple of books and later led me to develop a weekly column dealing with wildlife and with nature as a whole. As a result of shooting off my mouth in one of these columns, in which I deplored the lack of some public facility to care for wild strays, two baby raccoons were offered to me by a neighbor. Alvin Adams found the two starving coonlets on the edge of a local highway, a few miles from the Ontario town of Coboconk. Did I want to take care of them? What could we do but say yes?

Reluctantly, we accepted the waifs and with trepidation bought feeding bottles and two human infant nipples as well as a supply of baby formula and Pablum cereal at the village store. By trial and error we fed the suckling raccoons during the weekend, then realized that we must return to our Toronto apartment—which had an edict against pets of any

*See R. D. Lawrence, *The North Runner*. New York: Holt, Rinehart and Winston, 1979.

kind—with two rambunctious baby coons who didn't even have a proper cage in which they could be confined while we were both at work.

Before this I had raised a baby beaver, Paddy,* who furnished me endless hours of travail and pleasure in about equal doses and who was in the end allowed to return to his own world; but that adventure was different in that I was living in the wilderness at the time and my only neighbors were the other animals of the forest. Paddy taught me that the raising of wild strays was far from easy and now, operating under difficult circumstances, Joan and I were faced with two such foundlings.

In any event, Coby and Conk, as we named them in honor of Coboconk, the town near which they were found, appeared to thrive under the rather unusual conditions in which they lived. They were commuters, spending their weekdays in a city high rise and their weekends in the wilderness, two mischievous beings, one male, Conk, the other female, Coby, whose outlook was opportunistically simple: given food, affection, and lots of things to play with, they were content and not too destructive.

Never having learned that discretion is the better part of valor, I wrote about our two forest sprites in my column, and soon after its publication other strays began to arrive on our doorstep: a red-tailed hawk juvenile, named Maggot for reasons which will be made clear later; a young skunk; and two more raccoons, both young, but older than Coby and Conk. These had been picked up as "pets" by well-meaning people, but, when they turned out to be "a damned nuisance and very destructive," their owners wished to dump them. So it went—young birds, a duck with a broken wing, another that had stepped into a forgotten leg-hold trap and

*See R. D. Lawrence, *Paddy*. New York: Alfred A. Knopf, 1977.

had had one leg mangled—which I had to amputate—and an assortment of mammals in a variety of conditions, some ill and incurable and therefore having to be put out of their misery, others injured, some too young to care for themselves.

As the number increased, so did our reputation. At first only individuals moved to pity for the strays would call me or bring their finds to office or home; later the police departments in our region, only too happy to have someone else handle rescue calls that were really outside the area of their responsibility, got into the habit of enlisting my help.

At first we were simply moved by pity for our wards and didn't realize that a definite pattern was forming; that we had, in fact, become the sanctuary, at least in our relatively immediate area, that I had been advocating in my writings. Unobtrusively, we became "hooked." We spent hundreds of dollars annually on varieties of foods suitable for the wide range of animals that we cared for; other hard-earned cash went into materials for the construction of suitable, but temporary, cages to house those animals that were in need of isolation for a time. Our lives became totally bound up with our self-imposed task: we were the servants of nature. And we loved every minute of it—well, almost. There were times when sadness was our reward, and frustration, and anger at humanity for its cruelty; but, on balance, there were more rewards than penalties.

One day, and I cannot recall when this was, so it must have happened gradually, I realized that I, as a biologist, was learning a great deal more than I had ever been taught during my studies. The first such major milestone occurred when I became aware that raccoons do not wash their food. Coby and Conk were responsible for my beginning to wonder about the fable that every book on mammals persists in carrying.

When the two baby raccoons were old enough to require exercise, I rigged up a playpen for them in the bathtub, putting in "islands" for them to climb on while they played in their two-inch-deep "lake." To my surprise, they *hated* the water, loathing it so much that they showed their displeasure by defecating into the nasty wet stuff every time that, with typical human stupidity, I insisted on putting them into it. Too cold, perhaps, said Joan. We made the liquid tepid and it received two more discharges of mushy feces; we made it warmer yet, about blood temperature, with the same result. The two raccoons wanted no part of the experiment.

Perhaps the bathtub was not natural enough? I smuggled in branches, a couple of rocks, and a section of tree trunk and with these made them a "natural" habitat. Negative. Well, maybe they would behave differently at the property? The next weekend I carried Coby and Conk, who were quite content to go for a portage through the woods draped around my neck and shoulders, and set them down on the edge of our beaver pond. They loved this, playing with sticks and stones and weeds, catching ants, one at a time, and crunching them daintily, but after I had watched them for half an hour during which they did not so much as wet a single toe, I thought I would introduce them to their "natural" medium. The moment I put one of them into the shallows, it scrambled right out again. I fed them peanuts, thinking that these, surely, would cause them to go to the water so that they could wash them, something they had never yet done, despite the fact that we always put a large bowl of water in their open cage at the apartment—a failure on their part which I put down to the unnatural way in which they were living.

Coby took her peanut, chomped it, removed the shell, and ate the nuts; Conk played with his a bit, rolling it between his hands, then copied his sister's behavior. After each had consumed with relish ten peanuts (I wouldn't feed them more for fear of upsetting their ever-touchy bowels) without going near the water, I began to wonder. It seemed reasonable to surmise that either the raccoons had developed a dislike for washing their food because of captivity or the animal's reputation for dunking its provender was unfounded.

Seeking to find the right answer, I booked a month off from work (where I was owed seven weeks in lieu of overtime pay) and set myself up in a tent on a tiny island in the center of the beaver pond. From here, using field glasses, resorting to tracking, and after many hours of continuous observation of fourteen wild raccoons, I was able to establish that these animals do not wash their food. This was a satisfying discovery, but it left me with another puzzle: How did the raccoon earn its clearly undeserved reputation? Its Latin name, *Lotor,* means "washer"; it was opined by some that the animal had to wet its food because it didn't have salivary glands, or, at least, that these were inefficient and incapable of lubricating its food; another theory maintained that the raccoon's throat was narrow and it had difficulty swallowing because of this. How these preposterous concepts ever got started is beyond me. I knew from experience with Coby and Conk that both young coons had efficient salivary glands that were capable of drooling all over my fingers when I offered them a tidbit; I also knew that they had absolutely no trouble swallowing even the driest foods, such as a slice of stale bread.

As so often happens to those who study the natural world, having found the answer to one question, it now became necessary to discover the answer to another one. The weeks passed without enlightenment and then, when two raccoons that had been pets arrived, I was intrigued to note that both of them insisted on dunking their food in the waterpot. That offered me a clue. On querying the people who had kept them, I learned that the two had spent practically all their time in a cage. Putting two and two together and making a rather wide intuitive leap, I decided that captive raccoons wash their food because they are frustrated. The animal is a hand hunter, that is to say, its front, handlike paws are forever on the go, feeling, seeking insect, frog, or crayfish, or whatever might be edible. Deprived of its need to hunt, which in hunting animals is, second to mating, the next most powerful influence in their lives, captive coons that are sated with readily provided food play at hunting in order to relieve their boredom and frustration, "losing" uneaten food in their water container and "catching" it again.

The two newcomers were not kept in a cage when they came into our care. They were already old enough to manage on their own if they had a mind to, and we gave them this choice. They weren't ready to go. Instead, spending much of their time in adjacent trees, they would come morning and evening to get food. When they had been with us, free, for three weeks, first one, then the other, abandoned the habit of washing food. This was the clincher.

A few weeks after I had published my findings in the newspaper column, I was gratified to learn that Desmond Morris (author of *The Naked Ape*) had reached the same conclusions in England. Writing about the aberrations that develop in captive animals, his theories were presented by *Life* magazine in November 1968, under the heading THE SHAME OF THE NAKED CAGE.

In the time that followed this discovery I have come to realize that despite the massive amount of biological *data* available today, there is still a great deal to be learned about the *personality* of life. This knowledge can only be gained by studying animals in their natural habitat; there, in the wilds of our world, biologist and layman alike can still hope to discover some of the many missing pieces of the puzzle that is true Nature.

Thoreau must have had this thought in mind when, in 1856, he wrote: "*I seek acquaintance with Nature,—to know her moods and manners. Primitive nature is most interesting to me . . .*"

One

After the frost has finally gone, and before the emergence of the biting flies, there is a period in the backwoods each spring that would seem to have been designed for the inspiration of poets and the contentment of lesser mortals. The wilderness, in the short time since the last ice crystals glistened under the sun's melting rays, has managed to dress itself in soft greens and the bright pastels of the year's first wild flowers. At this season the animals show themselves readily. It seems as though a moratorium between hunter and hunted has been declared. Of course, this is not exactly so, for the meat eaters must still eat, and the grass eaters are, as a result, in as much danger as at any other time. But the trucelike atmosphere is there nevertheless, even if human ignorance is unable to understand it.

A newly pupated butterfly will dare to land on one's hand, there to sit opening and closing its colorful, still-damp wings while its coiled proboscis straightens, probes gently at the salt on the skin, then curls up again neatly, like a soft watchspring. A snowshoe hare is likely to stand upright, "washing" its face with its front paws only a few feet away from an observer. Birds will land on branches that are just above one's head and sit there singing so lustily that it might be supposed they have come expressly to give one personal pleasure.

On such a day I sat in the open, filling my pipe, my buttocks cushioned by a lush carpet of moss, the gentle sunshine that soaked into my clothing dispelling the chill acquired while traveling through the somber world of trees. Musing over the lack of caution exhibited by the animals and birds I had encountered during my walk, and remembering the many other spring days during which I had witnessed the same nonchalance displayed by many of the denizens of the wilderness, I once again tried to find a reason for such uncharacteristic behavior. This was an old challenge, a puzzle over which I had already devoted more than twenty years; and I was as far from solving it at that moment as I had ever been. But on the premise that enlightenment could come at any time, I worried the problem once again.

It was about 9 A.M. The night before there had been a light frost, a cold invited by clear skies in which the winking of countless millions of stars drew me into the darkened forest adjacent to our property, there to walk while listening to the wilderness and absorbing some of its tranquillity. Now, some three hours after sunrise, the glittering rime was gone and the sky, again cloudless, was a flat plane of translucent blue under which the sunshine mixed its gold with the delicate greens that dominated the landscape. No one could help being at ease in such surroundings on such a day; even the knowledge that death had been indulged here many times between last night and this morning could not dampen the sense of inner repose that I felt as I put a match to the pipe bowl.

I lay back, resting on my elbows, and allowed my eyes to travel over the land. My world was fragrant with scent and sparkling with color; wherever I looked there was beauty and the rhythms of life unfolding as the backwoods responded to the season. The melodies of a dozen or more different species of birds blended like the instruments of an orchestra to produce a woodland sonata that would have been the envy of a human composer; bumblebee queens, the only members of last year's colonies to survive winter by hibernating, were out and about, buzzing huskily as they sought suitable shelters in which to build new nests of fibrous wax preparatory to raising another brood of honey-dew eaters; a soft breeze tousled the crowns of stately white pines and caused them to sway as though in time with the music of the birds; somewhere on an unseen beaver pond a loon was cackling madly, its ribald voice yet managing to blend into the overall effect. I believe I would have stretched flat on the moss and gone to sleep if movement in the hitherto empty sky had not attracted my gaze.

I quickly identified five turkey vultures that had intruded into the periphery of my vision. Hanging in the air, the majestic black scavengers glided on immobile wings, their motive force the thermals that rose from the warming ground. With infinite grace and consummate ease the otherwise ugly birds wheeled, dropped, and rose again as the capricious updrafts dictated, ill-omened fliers staging a macabre aerial ballet that immediately returned me to my earlier thoughts. Vultures survive on the leavings of the predators and upon the putrefying carcasses of animals that have died as a result of sickness, injury, or age. Their presence always serves to remind me of the unending contest that exists between the predator and its would-be prey.

With their extraordinarily keen eyes, vultures can see a dead mouse as readily as they can spot the remains of a moose; with their large, flared nostrils that lead to well-developed olfactory centers, the birds can actually smell carrion from some 150 feet up in the air—this according to a test that I had made a few years earlier.

The presence of the black scavengers over my head was proof enough that the moratorium I had fancied earlier was extremely wide of the mark. Turkey vultures are not hunters, their feet being unsuited for grasping and holding prey and their beaks lacking the rapine curve of the hawks and eagles. In order for them to survive, death must be engineered by means other than their own. Clearly, I thought, as I watched the ebon squadron overhead, the incautious behavior of the lesser animals must be prompted by influences powerful enough to subdue their inherent vigilance.

What could make prey animals so careless? Love, perhaps? Not love in human terms, of course, but the urge to breed that grips the species annually in due season and which affects a large number of animals during the spring. This *might* be the answer, but it was only a theory, incapable of proof.

I continued to watch the vultures, wondering why the big birds seemed to be keeping station over a place on the other side of the ridge on which I reclined. Knowing the country, I was able to judge that whatever was attracting them was located somewhere between my ridge and the next, probably in the slight valley that separated the two rocky spines that traveled toward the northwest and continued in that direction for about five miles. At its end, I knew, the ridge of stone began to merge with an area of flat, open country paved with pink quartz and gray granite slab, visible effects of the cataclysmal eruptions that formed the Cambrian Shield of North America some 500 million years ago.

If the vultures were seriously interested in that place, there almost certainly had to be carrion down below. I expected them to land at any time, but after five minutes, during which they continued to circle over the same general location, I began to wonder. Vultures do not hesitate to drop swiftly on a meal when they encounter it, if it is safe to do so. The hesitation of the five birds suggested that whatever held their interest was either still alive or being eaten by a predator.

As I debated the matter and then decided that I would leave my comfortable "couch" in order to go and see what it was that interested the birds, I heard a faint, bleating cry that appeared to issue from the area below the carrion seekers. Moments later the cry was repeated; it contained a note of anguish that touched me.

I scrambled to my feet and walked briskly in the direction of the second ridge. Before I was halfway there the calls became more frequent, carrying what seemed to be a tone of special urgency. The animal's voice sounded rather like that of a small calf, but not so husky, and I had no difficulty identifying the caller: it was a black bear cub, one of this year's young, who evidently was terrified or in pain—perhaps both.

As I walked, I wondered about the cause of the little bear's anguish, speculating that perhaps the cub had become separated from its mother and, seeing the vultures, was afraid of them. On reflection, I discarded this idea because the calls were being repeated from the same place, and even very young bear cubs are agile little creatures that can run and hide or climb a tree to safety. This one evidently could not, or would not, move from that one location. It also seemed strange that the little animal's mother was not near enough to go to its aid, for it is unusual for female bears to become separated from their nursing cubs.

Thinking in this vein, I realized belatedly that I would be wise to proceed with caution and to try to find out what the trouble was without revealing my presence. Female bears are extremely protective of their young, and I didn't want to find myself near this yelling cub if the angry mother was coming to its rescue. Indeed, inherent caution suggested a retreat from the locale, thus allowing nature to take her course; but curiosity and pity for the youngster urged me on, though I slowed my pace and took advantage of all the cover that I could find while keeping one eye open for tall, but skinny, trees that would offer me safety but would be too thin for a cranky sow bear to climb. So I kept going, stopping often to listen for the crashing of brush that would tell me that mama bruin was on the way. Presently I spared a glance upward. The five vultures had dropped lower and were circling in line astern, traveling in a sort of follow-the-leader spiral.

Moving through the trees until I was about one hundred yards away from the place where I was sure the cub was located, it occurred to me that the little bear might be yelling because its mother would not allow it to suckle while *she* was eating the carrion that interested the vultures, for bears dearly love rotting meat.

While I was descending the slope of the ridge, the vultures came down to treetop height and the cub's voice became much louder. Over the years I had heard many young bears calling their mothers when the sow was some little distance away from them; none of these had cried with such intensity, or with such obvious panic in their voices.

The vultures, seeing me getting closer to their target, climbed higher, but continued to spiral over the same place, patient and cautious, prepared, as always, to abandon their meal rather than fight for it.

Presently I entered the small valley I had been aiming for, an area comprising approximately twenty acres of heavily wooded land, the soil of which was well watered by the run-off that weeped all summer long from the two ridges that flanked the low place. Here was plenty of shelter; the forest floor was soft and damp, moss-covered for the most part, creating conditions that allowed for silent travel. But visibility was poor; a cranky female bear, using her nose and ears instead of her eyes, could detect me long before I saw her, or heard her approach. Should the sow be in my vicinity, the first I would know of her would be when she came charging out of concealment.

Caution again urged me to retreat. I might have done so then if the cub's cries had not started to become weaker. I stopped, listening intently for the passage of a heavy body through the underbrush; only the cub's voice reached me.

Moving from tree to tree and uncomfortably on the alert, I traveled about thirty more yards before being brought to a halt by the stench of carrion. The nauseatingly strong odor told me that there had to be a lot of dead meat ahead. I had missed the stench until the contours of the land caused me to detour and turned me to face into the breeze, which was evidently fanning over the rotten stuff. The cub continued to bleat; it was near now, but still concealed from me.

Some little distance ahead, just before the land began to rise at the second ridge, there was a cluster of large rocks, craggy, misshapen blocks of time-eroded granite that had once been belched out of the guts of the world. The cub seemed to be on the other side of this natural battlement—so did the reek of carrion.

I abandoned caution and started to move noisily, coughing loudly and deliberately now and then, wanting to alert the she bear if she happened to be nearby. My racket silenced the cub; indeed, the wilderness around me became quiet. Apart from some distant birdcalls, the sound of the breeze rustling the treetops, and my own, rather heavy, breathing, I could hear nothing. But the odor of carrion had become so strong that I was close to gagging.

Half a dozen strides took me to the rocks and fully into the carrion's miasma, making me loath to scramble up one of the craggy monoliths so that I could look over its top into the valley on the other side. Breathing as shallowly as possible, I made myself climb the four or five feet that would allow me a view below. At first I could see neither cub nor carrion because of the brush that crowded this side of the little valley, but as I continued to scrutinize the ground I noticed, about ten yards ahead, the partly concealed outlines of an adult bear, lying on its side, quite dead. I moved higher, breasted the top of the boulder, and descended partway down the other side. From here I could see more clearly. The she bear must have been dead for some days; she was putrefying visibly, her guts already penetrated by wriggling masses of yellowish maggots. Crouched beside this mound of rotting meat and pressing tight against the back of it, was the cub, an emaciated, woebegone little being that was trying to make itself inconspicuous by pushing hard against the putrid thing that had once been its mother.

I forgot the stench and my earlier worries as I climbed down the rocks and walked slowly toward the terrified cub. Speaking softly and in an even tone, I sought to offer reassurance to this orphan of the wild.

The cub was about the size of a small fox, but a little taller. When I was three or four steps away, it began to snarl, curling its lips to expose tiny milk fangs, one of which was snapped off just short of the gum.

Young wild animals, even very small ones, look and sound extremely fierce under circumstances such as the one in which the bear cub found itself. The uninitiated human is likely to draw back from them, fearing a lacerated hand and the possibility of infection. In reality the fuss that these orphans make is bluff: they are terrified, and the need for comfort and security is uppermost in their minds. Their inherent instinct for survival, goaded by their fear, prompts them to show a belligerence that they do not feel and are not capable of translating into an attack beyond, perhaps, a graze or two inflicted by claws and teeth. I had by this time some years of practice in the handling of wild animals and had managed, hitherto, to avoid being seriously bitten or scratched.

Stooping down and moving my left arm toward the bear's muzzle to occupy its neophyte fury, I reached with my right hand and grasped the cub firmly by the scruff of its scrawny neck. It wailed shrilly, almost like a human infant, but tried to growl at the same time as it felt itself lifted from the ground and swung swiftly through the air toward my body. The little animal was amazingly light, not much more than three pounds, I thought, as it traveled toward me.

Pressing it against my shirt and holding firmly, but not roughly, with both hands, I continued to speak to it in a soft voice while it gripped me tight with its little claws and sought to hide its head under my left armpit, its growls becoming weaker and soon changing to whimpers. I had been expecting such a reaction. Allowing its muzzle to force itself between my arm and chest, I began to stroke it, backing away from the putrefying carcass of its mother, then turning around and walking slowly toward the rocks, but swinging wide of them, so that I could walk out of the valley without having to grip the cub tightly while climbing the boulders.

When we cleared the low place and had breasted the slope of the ridge on which the vultures first drew my attention to the forest drama, the cub became quiet. By now it was close to shock, its body quivering uncontrollably, its claws gripping me even more tightly and causing their needle tips to prick my skin. I was in something of a dilemma: shock can kill a young, undernourished animal as surely as a bullet in the head; we were a good three miles from home and I had no means of giving the little orphan the quiet darkness that it needed in order to steady its panic. In addition to this, I thought that at least some of its claws had scratched my skin and drawn blood and the danger of serious infection that the scratches posed was not to be taken lightly, for of all sources of contamination in the wild, carrion is among the most dangerous. Yet I had to walk slowly or the cub would become even more terrified. I compromised by lengthening my stride, but maintaining a slow pace.

In this way, while stroking the orphan and continuing to speak to it, it took us an hour and a half to reach home. Never was I happier to see the buildings of the farm when we at last emerged from the trees and started across the clearing that separated forest from civilization.

The bear cub was still shivering as I entered the farmyard and was met by Joan, who had seen me from the kitchen window and had noted that I was carrying "something alive." Without disturbing the little bear, I explained briefly what had happened and asked my wife to fix up one of our medium-size cages and put it inside the barn. While she was doing this, working with one hand, I started to prepare formula, a mixture of Olac, a human infant supplement, and cereal Pablum. I was still trying to get things ready when Joan returned, telling me, as she took over the task of heating the food, that I stank like a manure pile and that I had filled the whole house with my smell. Now that she mentioned it, I too could smell myself, but then, I was closest to the source: the cub. Such had been my preoccupation that the ripe aroma had ceased to bother me.

Joan was all for bathing the cub then and there, but I dared not further aggravate the bear's fear by dunking him in warm water and scrubbing him with shampoo.

"We'll feed him, then settle him in his box in the barn. He'll sleep and, later, when he's out of shock, we'll see about cleaning him up."

Wrinkling her nose at both of us while she opened a window, Joan nodded agreement, but suggested that since I wasn't in shock, I could take a bath as soon as the bear cub was settled in.

Faced with caring for yet one more waif, Joan bustled over to the barn with two eight-ounce baby bottles full of formula, a bath towel, a thermos jug full of warm water, and sundry other toilet requisites, looking back over her shoulder with a twinkle in her eyes.

"Come on then, mister. The sooner we feed this monster the quicker you'll get to smell like yourself again!"

In the barn, nestled in Joan's towel-covered lap, the cub fought only a little when I introduced it to the bottle. Holding his carrion-matted head with one hand, the bottle in the other, I waited until Joan had squeezed the rubber teat to cause a drop of formula to dribble out, and then I put the nipple against the cub's lips. He tried to edge away, but, feeling his mouth wet and being close to dying of thirst, he licked his lips. After that it was plain sailing, and when the second bottle had been drained, he would have consumed at least another had he been allowed to do so.

While I added some dry hay to the box to make a deep bed for our ward, Joan, having lost all her squeamishness, was stroking the stinking little creature and speaking softly to it, in a continuous soliloquy. The cub, no doubt still hopeful of finding another full bottle of formula, was snuffling her hands, the towel, and anything else that was handy. Joan stopped talking to him, chuckled quietly, then spoke to me.

"I've got a name for him already! Listen to his nose, the way he snuffles. We'll call him that, Snuffles."

And so we did, and Snuffles lived up to his name, for in all the years that we shared his friendship, he never stopped snuffling for food.

Two

When he was put to bed inside his hay-filled "den," Snuffles promptly went to sleep, if sleep is the proper word to use to describe the torpor that seized him almost at once. Replete, sheltered by darkness, and comfortable within his lair, the little bear surrendered totally to exhaustion.

Malnutrition, dehydration, and fear, leading to shock, had brought the cub almost to the point of death by the time I found him. These things explained his apparent docility, for the token resistance that he put up when I lifted him from the ground was only a weak imitation of the kind of savagery that a healthy bear of the same age would have resorted to if faced with similar human advances.

Joan, whose experience hitherto had been confined to the smaller mammals, expressed surprise at the cub's tractability when she held him on her lap preparatory to his first feed, not realizing at first that Snuffles was in such a weakened condition. As we stood in the barn listening to the little bear's lethargic breathing, I knew that at least another twenty-four hours would have to pass before we could feel more or less sure that he would survive. All that we had done so far was to give him the nourishment and liquid that he so badly needed; the rest was now up to him; or, rather, to the physical and mental systems with which nature had endowed him. If he had not become the victim of illness during the time that he had been on his own, if malnutrition had not gone beyond the point of no return, and if the shock he had suffered did not prove to be irreversible, he probably would recover.

Back in the house, seeking to prepare my wife for the upset that would, I knew, follow should the cub die, I explained some of these things to her, particularly the possibly lethal effect of shock, a condition with which most people are nominally familiar but which few laymen really understand. Medically, shock is classified according to its causes. It can result following loss of blood, serious burns, fear, poisoning, surgical operation, and even from medical treatment. The condition produces a variety of extremely complex symptoms that seriously upset body functions. Basically, and at the risk of oversimplifying, what happens is that the walls of the capillary blood vessels become excessively porous and allow the blood's plasma proteins to escape through them; this reduces the blood flow; blood pressure drops and tends to continue falling, leading to irreversible loss of circulation, which ends in death. Treated in time, shock can be successfully alleviated by transfusions of blood plasma, which replace the lost proteins, but we had neither the equipment nor the facilities for such treatment; and even if we *had* been so equipped, it would have been necessary to secure the right plasma—in this case from another bear!

Trying to spare my wife from worry while preparing her for the worst, I downplayed the dangers that faced the cub and ended on a note of hope that I did not feel by telling her that if he survived the next twenty-four hours, he probably would be all right. As usual, I underestimated Joan's inner fortitude. Though tears filled her eyes when I ended the explanation, these were quickly replaced by a look of determination that told me that my wife had now taken the bear's survival as a personal challenge. If quiet care and great devotion could make a difference—and I believed they could—the bear cub would live.

Survival in the wild frequently depends upon personal determination, or strength of character, as a human would say. This mysterious *something* that is genetically programmed in greater or lesser quantity in every individual of every species is still largely beyond the reach of human understanding. I think of it as an "X quality" implanted in the brain and nervous system of each animal while it is still in its mother's uterus. Some organisms seem to have more of the quality than others; some, after birth, and through a combination of parental care and positive environmental reinforcement, develop their *X* quality, building upon the foundations that nature supplied. Leaders are made of this; so are survivors. Rather naturally, the size and physical condition of the body tends to add to, or subtract from, the chances of survival, though this is not invariably the case: some relatively puny individuals who are particularly strong-willed make the grade regardless, proving, it seems to me, that muscle is the servant of wit, and that while brawn is an excellent and desirable commodity, it is soon wasted within the scheme of nature if it does not have a keen and determined brain to guide its exertions.

Bearing these things in mind, and having done what we could for Snuffles, I tried to be philosophical about his chances, seeking to implant in Joan's mind a similar acceptance, yet knowing that I was wasting the effort. Indeed, no sooner had I finished speaking than my wife picked up a folding stool and went with it to the barn to squat beside the cub's shelter. There, patient and hopeful, she listened to the little bear's breathing while now and then reaching inside the den to stroke him gently, her entire mien suggesting that she was seeking to will life into the orphan. Recognizing her symptoms, I told her that I would take care of chores and would prepare the meals.

Snuffles, in the meantime, remained oblivious. He slept, he snored, he even snuffled in his dreams. If he wasn't in a coma, he was in the nearest thing to it, causing me to lose more and more hope as the hours dragged by. In vain did I seek to relieve Joan at the patient's bedside during that evening. She did consent to go back to the house to eat her supper, but she did less than justice to my culinary efforts.

With a few breaks for coffee and bathroom facilities, and some snatched periods of sleep while I stood watch over Snuffles, Joan spent twenty-seven hours in vigil. Then, at suppertime the next day, just as I was about to go to the barn to *order* her to come in and eat, she burst into the house with joy radiating from every part of her being.

"Quick! Let's get some formula ready. He's *awake*!"

Having delivered her terse but happy statement, she wheeled around and ran back to the barn, causing me to realize that the use of the plural *let's* was to be translated into the singular *you*. I was, of course, pleased with her news, but I made myself accept it with caution. For Joan it was enough that her ward had awakened after sleeping more than twenty-four hours; for me, the cub's future remained uncertain. But I hurried as I mixed Olac and Pablum and filled two eight-ounce bottles.

When I carried the food into the barn I was somewhat surprised to see Snuffles lying on his back in Joan's lap, all four legs in the air and his mouth fixed upon one of my wife's thumbs, on which he was sucking avidly, causing his benefactress some pain, judging by the grimaces she made whenever the cub chomped too hard on her digit, no doubt rasping her skin against his rough baby teeth.

It was now late evening and therefore gloomy inside the barn, which had never been hooked up to electric power. But whereas the lack of illumination was a nuisance in that I couldn't readily see what I was doing, it was also a blessing, because the cub was clearly more at ease in conditions of subdued light and showed no signs of fear. He cringed a little when I reached down with a full bottle of formula and he growled faintly once, but as soon as the rubber teat touched his lips he seized it and began to suck greedily, letting his body flop down in Joan's lap while his front paws came up and gripped the glass container between them. The eight ounces of food disappeared rapidly and Snuffles resisted strongly when I began to ease the nipple out of his mouth. When this was at last removed, and during the few seconds that it took to switch the empty for a full bottle and put it within his reach, the cub became almost frantic, making small whimpering sounds and reaching with his front paws, so that they were ready to receive the second helping. Like its predecessor, this feed was gulped down rapidly, even if the cub paused a few times to get its breath and to loudly expel trapped air from his stomach.

Now we found ourselves faced by a familiar quandary. Should we allow the bear to ingest another eight ounces of food, or should we not? Starving animals will gorge themselves if they can, as we well knew from past experience. In young mammals the result of such injudicious feeding is usually followed by severe stomach upsets, in some cases serious enough to cause death. Yet it is always hard to gauge the needs of a starving animal, particularly when one does not know if it is being greedy or if it really requires more food. As we had so often done before, we compromised by allowing the cub another four ounces of formula. Happily, this seemed to satisfy him and, as we discovered the next day, did him no harm. More important, the extra feed restored his good humor.

The reek of carrion that filled the barn reminded us again that the bear cub was filthy, but I still felt it was too soon to subject him to the necessarily ungentle business of cleansing him thoroughly in the bathtub. In the wild, a mother bear, like most females of animal species, cleans her cubs by licking them thoroughly, her damp tongue washing off the outer, or guard, hairs, but leaving the soft, fine, inner wool relatively dry. This prevents the protective oils, which coat the inner wool and make the animal virtually waterproof, from being removed. Since it is the long outer hairs that usually become dirty, thus shielding the wool from debris, and because the mother's rough tongue is such an effective instrument with which to scour the matted filth, the young animal is rarely seen wearing a dirty suit.

As a rule, we did not bathe the orphans that came into our care unless they were in deplorable condition. Usually we limited their toilet to a good swabbing with a damp sponge followed by regular brushing, a procedure that came close to imitating a wild mother's care and that did not soak the underfur and remove the protective oils. But there were exceptions, and Snuffles was one of these. We could not remove the crusted carrion from his body without a thorough soaking and the application of baby shampoo. This would not harm him unduly insofar as he was not going to be subjected to the risk of rain and the body oils would eventually reanoint his underfur, but I felt that forcing him into a tub and holding on tightly against his natural anxiety to get out of the water might upset him and cause us to lose his trust. Later, when he was more settled and had given us his full trust, we could risk it. Now, I compromised by getting a basin of warm water, a sponge, some baby shampoo, and a heap of old towels.

While Joan continued to hold him, I soaked a sponge, applied to it a little shampoo and began to swab, expecting some violent reaction from Snuffles. Fortuitously, he either confused the sponge with his mother's tongue, or he liked the feeling of the warm water and the relief that this must have provided him. In any event, we managed to sponge him fairly well, rinsing him bit by bit and then rubbing him vigorously with the towels. There was no doubt that he really enjoyed the rubbing. He actually moaned with delight, especially when I did his belly and private places while he leaned forward and watched my every action with what I am sure was a gleam of appreciation in his shiny black eyes.

When he was all done and smelling almost like a bear instead of like a piece of rotten meat, Joan picked him up under the armpits and hugged him to her, whereupon Snuffles nestled his head under her chin as he gripped her gently with all his paws, the while doing a small piddle down the front of her blouse.

By this point my wife was accustomed to such accidents, treating them quite coolly as she admonished the offender in a tone of endearment. Indeed, discussing such mishaps after her first year tending to our wild orphans, Joan told me in all seriousness that young wild animals were less messy than human babies because they developed at a faster rate and were soon able to go to the toilet on their own. Since Joan had amassed considerable experience while caring for a brother very much her junior, I bowed to her superior knowledge. Certainly, as she pointed out, cleaning up a young wildling was "duck soup" compared to changing and washing dirty diapers on a regular basis. As for me, after having cared for nineteen cows, a number of hogs, chickens, and dogs during my years as a backwoods homesteader (not to mention cleaning out hundreds of latrine pails while I was an army recruit), the body wastes of animals were a matter of interest rather than of disgust. Much valuable information can be had by examining the feces of wild animals, a practice in which I still indulge when out in the field, at times even taking small samples in suitable containers for later microscopic analysis. From such examinations it is possible to determine the diet, condition, and prey species taken by a predator; and even grass eaters yield similar dietary information.

Apart from soiling and wetting on a regular, normal basis as the need arises, mammals are prone to emergency discharges when they become nervous or actually afraid. This occurs in *all* the species, even our own; it is part of the mechanics of survival designed by nature to ready the body for escape. Undischarged wastes are a hindrance to sudden movement during an emergency and are therefore expelled even as an animal starts to flee, or when it is suddenly cornered and seeking escape. Knowing these things, we tried to reassure each newcomer as quickly as possible and we were always careful to avoid sudden actions that might startle our charges. Calm movement indicates neutrality to a wild animal; slow, stealthy progress is identified as the stalk of a predator; and sudden action is interpreted as the commencement of an actual attack.

The evening of the cub's accident on Joan's blouse, I suggested that because Snuffles was so docile and trusting, we might now take him to the house, to give him a chance to explore it and to learn that it was as much a part of his new territory as the barn. Joan, who wanted to change her wet garment, readily agreed.

It was almost dark when we crossed the space between

the two buildings and entered our lighted dwelling, but the cub did not react to the sudden change. On the way over he had continued to nestle against Joan's neck, remaining placid; when we stepped into the kitchen he raised his head and began sniffing the plethora of odors that his nose immediately encountered, his behavior indicating interest rather than apprehension, a circumstance that didn't surprise me in that the scent of cooking and stored food were to him easily detectable, even if we could not smell them.

Joan handed Snuffles to me, an exchange that was effected without fuss, except that the bear elected to settle himself high up my shoulder, from where he continued to sniff and to allow his gaze to roam over all the interesting things that he was encountering for the first time. I decided to weigh him, using a balance that we kept in the kitchen for just such a purpose. Our usual procedure was to step on the scales while holding the animal and then deduct our own weight from the total, though such a technique was only required with the larger mammals, the small ones being placed on Joan's kitchen scales, for which I had fashioned a small, wire-mesh basket to hold the animal. My initial guess turned out to be close to the mark; the little bear scaled three pounds two ounces. This meant that he had probably lost almost fifty percent of his weight between the time that his mother died and when I found him. It also meant that we must fatten him up as soon as possible.

When Joan returned from upstairs wearing a smock that she reserved for handling young animals, I gave Snuffles to her once again and, taking a battery lantern, returned to the barn to change the soiled bedding and to mix in the new straw a liberal quantity of antivermin powder, for we had not managed to rid Snuffles of all his tiny tenants.

It says a great deal for my wife's devotion to our "critters," as we often referred to them, that she accepted with better than good grace the lodgers with which most wild strays arrived. These buglets included wood ticks, fleas, lice, and warble-fly larva, the last being harmless to humans, yet eliciting the greatest disgust from Joan. In truth, these larva are quite repulsive creatures! The female flies of this species are fast-flying, hairy insects that are about half an inch long and are parasitic on most wild mammals and on cattle. The warble of North America lays her tiny, sticky eggs either singly or in a series on the hairs of her victims, and when the larva hatch they penetrate the skin of their "host." At birth they are very small, but once under the skin of their victim they grow big, eventually becoming plump, chocolate-brown carnivorous pests an inch long and a quarter of an inch wide that are enclosed in a cyst formed by the tissues of the animal they have invaded. Because they must breathe, these revolting creatures cut a hole in the skin of their unwilling carrier, out of which they eventually squeeze when they are ready to pupate in the ground. Eggs are usually laid during summer and autumn, and the larva spend the winter and spring inside their victims, emerging during the warmer weather. Although most reference works on insects stress that the warble flies are pests of the larger mammals, such as cows, deer, and moose, no warm-blooded animal is safe from this fly's depredations; even mice, chipmunks, and squirrels are invaded, but whereas the larger animals, though they may suffer weight loss, are not greatly affected by the grubs, the small ones may die as a result of a serious invasion. Snuffles, because of his age and the season, was not housing warble grubs, but it would be necessary to check him continually as the weather warmed, which was something that we had to do for all our animals.

Working in the barn, I sought to estimate the cub's age, reviewing my knowledge of bears. Seeking to pin down the date of birth of a stray animal is always difficult, because mating and gestation times are irregular within each species. At best one can only hope to set an approximate fertilization time and then calculate a borderline birth date. Black bears mate during early summer, but precisely when this occurs in each instance remains a matter between the individual female and the male that happens to meet up with her when she is ready for his company.

The connubial relationship lasts about one month; during this period the female is fertilized, but whether the first or last union produced the desired result (or any other union in between), is something that only the she bear would know. Then again, early summer can stretch from June 21 to mid-July, according to human calendars, but this is an arbitrary time scale as far as the bears are concerned. But one must start *somewhere*, so I chose June 30 as the date on which the cub's mother received the male's sperm in one or more of her ova. The next step was to decide on a gestation period, and once again, it was not possible to be definite. It is generally considered that female bears carry the developing embryo for a period of twenty-eight to thirty weeks, but as is the case with humans, this time can be shortened or extended by factors affecting the individual female. In this case, I chose twenty-nine weeks. If Snuffles was engendered on June 30, he could be expected to have come into the world some seven months later, or about the end of January.

Baby bears are born during the female's denning period, a long sleep that is not true hibernation in that her body temperature remains normal and her breathing, though slow and lethargic, continues at some five complete respirations a minute, compared with a true hibernating animal, such as the groundhog, whose temperature drops to as low as 47 degrees Fahrenheit while its pulse and respiration rates almost cease, with only the merest trickle of air entering and leaving its lungs. But whereas the groundhog is in such a profound sleep that it is insensible to touch, the bear is at times semiconscious and aware of affairs within or outside of its den, becoming wide awake once in a while and likely to react violently if intruded upon.

Labor pains arouse the she bear, who sees to the biting of her offspring's umbilical cord, cleans up the afterbirth by eating it, licks her cub (or cubs) clean, then drifts off to sleep again, leaving the young to find her milk-full dugs and there to suckle at will.

Bear cubs are tiny creatures when first born, measuring no more than nine inches in length and weighing between six and eight *ounces*, which is only about one four-hundredths of the average she bear's weight. The cubs are blind, hairless, toothless, hard of hearing, and have short, stumpy legs and a ridiculously small tail; indeed, both the legs and the ears are only just obvious, the former barely helping the cub to wriggle toward its mother's breast and the latter virtually unable to hear even the nearest sounds.

Thus, while the outside temperature may be thirty or forty degrees below zero, the cubs thrive, drinking at will as they nestle against their mother, further sheltered by a bed of grasses, evergreen needles, bark, and leaves that she has dragged into a cave or rocky hollow. Then, after about forty days and nights, the eyes of the cubs open and they begin to develop their first teeth. Now their legs and ears have grown and their bodies are covered with soft, black down; they have also gained weight, scaling some two pounds, and they are about one foot long.

In the spring the mother leaves the den and leads her cubs into the new world, going slowly so that her toddlers can keep up with her as they learn to walk properly. The cubs now weigh about four pounds. This first vernal outing takes place according to the latitude of the bear's home range: where winter lingers, the bears emerge in late March or mid-April; in milder regions they come out earlier.

Aided by these details, I concluded that Snuffles was a little more than three months old when rescued on May 7. Had I found him soon after he had become an orphan, his size and weight would have helped me establish his age with more accuracy.

Birth dates for young animals are not the important social occasions celebrated by our own species, and my concern over Snuffles's arrival in the world was not prompted by sentimentality but by the need to know what, and how much, to feed him once he recovered his health. A brief examination of his teeth, undertaken when introducing him to the bottle, suggested that he had been driven to chewing debris and whatever bits of edible matter he could find when acute hunger forced him to explore the area around his dead mother. Both his upper canines were broken; the one that I had first noticed was snapped off right at the gum line and the other one about halfway up, indicating that in his ravenous despair he had chewed on branches and, probably, rocks.

My estimate of his age, awareness of what a bear cub should approximately weigh by this stage in its life, and the broken teeth combined to tell me that Snuffles could already eat some solids unaided, but needed the solace of sucking from the bottle—as a substitute for his mother's dugs. This information determined the form and substance of his diet.

The evening of his arrival, while Joan sat in vigil in the barn, I wondered about the death of the she bear, even debating the wisdom of returning to the scene in the morning to seek to discover the cause of her death. Adult bears can certainly die of disease, but they are a pretty healthy lot, by and large, so I was curious. Yet the prospect of again facing the mound of rotting tissue, and poking about in it, was unpleasant enough to put a damper on my curiosity. Perhaps I might have undertaken the task if I had really felt that it was still possible to establish the cause of death, but by then the state of the body was too far gone to show evidence of disease, at least with the limited resources that were at my disposal. Had she been shot, which I suspected was the case, even the wound would not be discernible by now. Putting the matter to one side, I determined to wait until the carrion feeders had done with the remains, then to go back to see if I could find any bullet holes in those bones not chewed to pieces.

I was also curious about the number of cubs that the bear had produced in January. Young female bears, mating for the first time, usually have only one cub; older females may have twins, triplets, or even as many as five young, though quintuplets are fairly rare. The size of the dead bear suggested that she was a mature animal, but whether she had produced more than one cub and the others had died or been killed by predators was impossible to tell. I was reasonably satisfied that if she had borne more than one offspring, they, like Snuffles, would have remained in the general area and I would have seen or heard them, if they had survived.

When I finished arranging our new ward's bed and returned to the house, it was to find Joan feeding the cub some of our home-produced maple syrup. The bear was again lying on his back in her lap, a position and place that were to become his favorites while he was young, sucking syrup from my wife's baster, one of those clear plastic tubes with a narrow nozzle and a rubber bulb at the handle end with which turkey and other roasting meats are squirted with hot fat.

If there is such a thing as bear heaven, Snuffles was in it at that moment. His small, handlike front paws gripped the syringe, his rounded lips were glued around the nozzle, and he sucked noisily while Joan squeezed the rubber bulb. Delight peeped shyly from his half-closed eyes; his shiny black nose wiggled from side to side as he doubled his pleasure by sniffing as well as tasting the delectable sweetness. The syringe was quite large, about seven inches long by one and a half in diameter, as I recall, and Joan evidently had filled it. If Snuffles drank the lot, we would have a sick cub on our hands. I cautioned my wife about the possible consequences of feeding him such rich food in his present condition and she immediately realized the danger and tried to remove the tube from her ward's mouth, whereupon Snuffles gave way to a terrible and noisy tantrum.

During the few seconds that passed before I hurried to Joan's aid, I could not help marveling at the speed with which Snuffles altered character. At one moment he was the picture of infantile bliss, at the next he transformed himself into a furious, and seemingly already spoiled, brat. When he felt the end of the baster begin to slip from between his lips, he gripped the syringe tightly with his forepaws, his lids opened wide, and the spark of anger lent a malevolent look to his previously soft, warm eyes. As Joan continued to pull the tube away, the cub raised both back feet so as to get a better grip on the body of the syringe, digging his claws into one of Joan's hands as he did so. By the time I took the three or four steps that brought me near enough to reach the bear, he had literally wrapped himself around the container and my wife's hands.

Joan, though in obvious pain from the digging claws, tried to remain benevolent, speaking to the cub in a gentle voice and asking him to "please let go," but Snuffles, even had he been able to understand her every word, was in no mood to be parted from the syrup. He had never tasted the stuff before, but now that he had found this elixir, his every action bespoke his intentions of fighting to keep it. Seeing that Joan was already bleeding from two fairly deep scratches on her right hand, I went to her rescue.

Securing a good grip on the back of the cub's neck with one hand, I firmly grasped both his back feet in the other and started to pull him upward. He screamed a cry of primitive rage quite unlike his plaintive wailing of yesterday; but he did what I expected him to do: caught from two sides and feeling himself helpless as he was hoisted heavenward, he released his front paws from the baster and sought to attack one of the hands that seized him, cutting off his shriek and turning it into a growl that despite having baby overtones managed to sound quite ferocious. While he was reacting in this way, I lifted him off Joan's lap and let go of his back legs, leaving him to dangle in midair. But he didn't give up his fight, so I shook him a time or two, speaking gruffly, seeking to imprint on his mind some human discipline. Perhaps my deep tones had an effect, but I am sure that the shaking, by causing him to feel completely helpless, exercised the major influence. Now he stopped growling and began to whimper pathetically.

Joan, ignoring the blood that dripped from the back of her hand, immediately went to her ward's rescue. Almost leaping from her chair, she wrapped both arms around Snuffles and hugged him to her so that, perforce, I had to let go. As soon as he was enfolded in that maternal embrace, the recalcitrant little bear snuggled into Joan, again placing his muzzle tight against her neck, there to whimper softly for a few moments while his head and back were stroked gently. Seconds later he was quiet and completely docile; the altercation was over. That is to say, it was over as far as Joan and Snuffles were concerned, but now it began anew for a quite different reason and with me as the central villain. Without raising her voice, Joan made it clear that she considered me a brute.

It must not be supposed that my wife and I were always engaging in arguments as a result of our different approaches to the animals that we cared for. On the contrary, we *never* seriously argued or became cross with each other; but we did indulge in a sort of game. Joan enjoyed shielding her wards from my supposed roughness. I equally enjoyed playing the part of the hardened biologist who would not allow the wild ones to be spoiled by her softness. So we teased each other, often pretending anger that we did not feel, especially if other people were present. I am not sure why we did this, or even when it started, but it did help us at those times when problems or emotional upsets cropped up, which under the circumstances occurred with regularity. In a sense, Joan was my pupil. She had not been especially interested in wildlife before meeting me, not having really been introduced to it, but from the first time that she accompanied me on a field trip she came under the spell of the wild. And she was a natural with animals, probably because she never became angry with them and was always serene, her aura of peace and tranquillity communicating itself to the forest folk. As the years passed, she absorbed knowledge rapidly and readily, and although she was not a reader and disliked having to plod through my textbooks, she more than made up for this by asking innumerable questions, listening to the answers, and remembering them.

While Joan was chiding me for my behavior toward Snuffles, our latest orphan, with typical juvenile changeability, stretched himself in her arms and began to play gently with a necklace of pink beads that she was wearing, assuming a pose of injured innocence. I took this opportunity to become serious, explaining to Joan that the young bear was still in poor condition and liable to sudden bowel upsets if he was allowed too many tidbits or too much of a particularly rich treat.

It is always easy to spoil a young, cuddly wild animal; they are so appealing. And so smart! They learn speedily that they can charm extra goodies from a human by posing pathetically, or looking coy, or even, as is common with the bear clan, resorting to out-and-out clowning, perhaps standing on their hind legs and extending their open front paws, or doing a little dance. Anything, in fact, that produces a pleasurable or amused reaction in a human is quickly stored in the box of tricks that each hides somewhere within the deep recesses of its brain, from there to be used when occasion demands, each new self-taught "trick" adding to what eventually becomes a quite extensive repertoire.

Humans honestly believe that they train their pets, whether the animal in question is a dog, a cat, a hamster, or a canary. As far as I am concerned, it is the animal that trains the human. When Fido realizes that by rolling on his back he gets an extra cookie, he will do so every time he believes his master is in a receptive mood; and when visitors arrive and master wants to show off his smart dog, the latter does a *real* number, knowing full well that several treats will be forthcoming under these circumstances.

The reason for this is that the majority of mammals are opportunists who are quick to read all the signs that point to food as well as all the signs that lead to danger. Because food in the wild, and for domestic animals as well, is such an important part of life, animals seek it at every opportunity, readily adapting themselves to those conditions that result in a meal. This means that an animal soon learns the craft of hunting, or the tricks of eating the best grass, while keeping eyes, ears, and nose on the alert for meat eaters. It also learns to change its eating habits in accordance with the food supply. (There are specialist animals, of course, who restrict themselves to particular foods, but these are a minority when compared to the opportunists.) In this way, hares and squirrels and chipmunks, who are not thought to be carnivorous, will eat meat when they can get it, sometimes going so far as killing it for themselves; the beaver, long considered something of a specialist in that it eats tree bark in fall, winter, and spring, consumes water and land plants when these are in season, also taking animal protein when it is available. It is all a matter of opportunity and of the state of the appetite.

Under these circumstances, when an animal discovers that it can get what are, in fact, appetitive luxuries from a human, it seeks to repeat those things that obviously produce the desired results. If one laughs and claps at a zoo bear when it does its party trick and then one rewards it with peanuts or buns, the animal will continue to dance, or wave, or beg with both hands, just as long as the visitors are there to respond to it.

An hour before bedtime on the second day of the bear's arrival at our farm, I beat two eggs into enough cereal Pablum to make a pap into which was added a quantity of raisins and half a banana, the whole being presented to Snuffles in a metal plate, for we were anxious to learn if he could be induced to feed himself in this way. We need not have worried! All it took to get him to stick his face into the mush was one whiff of the mixture. He attacked the food with gluttonous energy, planting both front paws into the dish while he ate.

The plateful was noisily absorbed in short order, after which the cub licked the container spotlessly clean, then proceeded to lick his paws, chest, and the floor, completing the cleanup so effectively that the pan of warm water containing a sponge that Joan had prepared was hardly needed.

The way matters developed that night, my work in the barn was wasted. Joan insisted that Snuffles must sleep in our bedroom on the premise that he probably would feel lonely and insecure if left to his own devices in the barn. This idea had not occurred to me, but I conceded that my wife was probably correct in her view, and, in any event, if *she* did not object to sleeping beside a somewhat odoriferous baby bear, I certainly would not.

To this end I went out to the storage shed where most of our unused nesting boxes were kept and found one that was three feet by three feet on the inside and came equipped with a lid. Into this I placed a good quantity of clean hay, again mixed with antivermin powder, and returned to the house.

Joan carried Snuffles upstairs to our bedroom while I followed with the box, which I placed between our twin beds. My wife laid down her ward with as much tenderness as a human mother would settle her infant in its cradle, stroked the cub for a moment or two, then stood back to watch him. The cause of all this fuss and attention stretched luxuriously, yawned once, turned himself into a black ball, and promptly went to sleep.

Three

Snuffles began his third day at the farm by rising early and leaving our bedroom without disturbing our slumbers, a quiet exit motivated not by consideration for his rescuers, but rather because Nature has endowed the bear clan with well-padded feet that carry their owners through the wilderness with the stealth of a hunting leopard. With feet like his, Snuffles could walk over our carpets as silently as thistledown drifts through air. When he reached the ground floor, however, and especially when he found the kitchen, events took a different turn.

Joan and I, blissfully unaware of our ward's absence, emerged from sleep into a state of partial somnambulism when a series of stentorian crashes shattered the quiet of dawn. Rubbing my eyes and throwing back the covers, I was slowly rising when further crashes lent urgency to my limbs, for now I was convinced that some part of our house was collapsing. I ran, noting as I headed for the stairs that Joan was sitting up and reaching for a housecoat.

I arrived at the scene of the action in time to witness a number of bottles and cans tumble out of a food cupboard located above the stove, on the enameled top of which the cascading containers were impacting.

Somewhere inside the human head there is a marvelous center that is able to register simultaneously, with nanosecond speed, a multiplicity of scattered details that escape the conscious levels of the brain. In a flash, with photographic clarity, one is able to absorb by this means an entire panorama while being consciously motivated by other stimuli. It was thus with me at 6:20 A.M. on Day Three of Snuffles, or, as I was to refer to it later, as the start of the Post Peace Period at North Star Farm.

My arrival in the kitchen coincided with the ejection from the cupboard of half a gallon of vinegar in a glass jug, a one-pound jar of pickled beets, a can of blackberry-and-apple jam, and a can of baking powder. The glass containers hit the appliance top and smashed, spraying acetic acid and beetroot over the countertop, the walls, the floor, and the stove; the jam and baking powder cans hit the already considerable mess on the electric elements, bounced elegantly, then fell to the floor, where each spat off its lid and fanned its contents far and wide. Even so, vinegar, beetroot, jam, and baking powder were mere accents, daubs of finishing shades added to the elaborate canvas. The whole scene was chaotically superb: a landscape of shrieking color combining to form a surrealist's nightmare.

The cupboard had been built to Joan's specifications. It was high off the ground to discourage our more active wards (*pre*-Snuffles) from climbing to explore it and it was wide and deep with plenty of room between its three shelves for tall containers. Helping to clear the supper table the evening before, I had noted that the cabinet was full almost to bursting. Now it was practically empty of provisions and in place of the rightful contents, squatting on the bottom shelf—the middle one having been tossed out earlier to cartwheel across the floor—was our hog-happy little bear, too engrossed in slurping up the contents of a two-pound can of strawberry jam to do more than spare me one quick and casual glance. The bottom shelf was liberally spread with a mixture of substances, some sticky and viscous, others clear and runny, that combined to drip off the shelf and down onto the stove top. Snuffles himself was well plastered with molasses, maple syrup, icing sugar, and what I took to be marmalade, in addition to other unidentifiable foodstuffs.

When Joan entered her kitchen she stopped, remained stock-still for some seconds while her eyes roamed over the chaos, and then, catching sight of the cause of all the trouble as he switched his attention from the strawberry jam to the ooze covering the shelf, burst out laughing. But that hardly describes her mirth! She cackled, she hooted, and she shrieked, clutching her sides, tears starting from her eyes. Hysteria, I thought, wondering if I should slap her face to bring her out of it. At that moment Snuffles stuck his anointed head out of the cupboard to peer toward the source of the new noise; his action and Joan's reaction to the mayhem committed in her kitchen became too much for my own sense of humor. As I too succumbed, the rise in the level of madcap merriment caused the cub to look upon us with utter amazement, his bewildered expression becoming so comical that it reduced us to a state of collapse.

Eventually, exhausted and aching, we stopped gibbering like lunatics. Drying my eyes with the heels of my hands and wiping the moisture on the seat of my underpants (my sole item of apparel), I advanced to the cupboard, looked the bear over critically, and announced to my wife that now he *must* be bathed, regardless of his reactions to the indignities that this involved. But several problems presented themselves before this could be done. Snuffles had no doubt already consumed more rich things than were good for him; thus it was necessary to put a stop to his gluttony immediately; the mess that he had created must be quickly cleared up if we were to inhabit the kitchen and get some breakfast; I was too scantily clad to enter into a wrestling match with even a *small* bear; the bath and its attendant paraphernalia had to be prepared right away. So, seeking to deal with each problem in sequence, I suggested to Joan that she should remove Snuffles from the cupboard and hold him while I went to cover myself with some stout work clothes, then, while she was continuing to pacify our ward, I would prepare his bath.

My wife, though agreeing in principle with my proposal, did not take kindly to the first part of it. Fixing me with a hawklike stare, she shook her head, this simple action managing to convey her veto more firmly than her spoken refusal.

"*No way!* Since you're practically naked, *you* get him down and *I'll* run the bath, because if you think I'm going to put that filthy little pig against my new housecoat, you're out of your mind!"

Having delivered herself of this succinct homily, Joan smiled sweetly at Snuffles and headed for the bathroom, leaving me in no doubt that she had insulted the cub in order to make her point to me and not because she was angry with him.

After her departure Snuffles and I regarded each other for a moment or two. The bear cub's gaze was quizzical, mine, judging by the way I felt, must have reflected a mixture of anxiety and distaste—the former because I wasn't all that knocked out by the concept of exposing my bare skin to twenty sharp claws, the latter because the prospect of clutching to my nakedness that tacky, slimy body was far from appealing. I was thinking about going upstairs to put on a work shirt when Snuffles decided that it was time for him to come down from his redoubt, perhaps with some idea of probing all the other cupboards that the kitchen contained.

The bath was running, but Joan had not yet called to say that all was ready. So I had a choice; either I could get a shirt and thus allow the cub to do some more prowling on his own, or I must grab him now and hold him until everything was ready for his big wash. I voted to go for the shirt; but before I took the first step, Snuffles slid to the verge of the shelf, turned himself around so that his bottom now aimed outward, and then eased himself over, miscalculating his move so that he almost fell and only just managed to hook his front paws on the very edge of the wood.

He became frightened and yelled; and each time he opened his mouth, his claws slipped closer to the rim. Reflexively responding to his fear and to his predicament, I reached up and plucked him off the shelf, his back turned toward me; then I swung him around with the intention of holding him at arm's length. But I misjudged the extent of his reach and his plump front paws, all ten claws spread, were able to get hold of my shoulders, one on each side of my neck. He began to draw himself toward me by digging with his hooks and pulling. I opted to give rather than to resist, in this way allowing myself to be smeared by the mixture that soiled his coat, but preventing his claws from making "train tracks" on my hide.

Despite my reluctance to clutch him to me, I was intrigued by his behavior when, on feeling himself securely held, he cuddled into me, resting his chin on my shoulder and allowing his weight to settle against my hands and arms, his nervousness of moments before vanishing instantly. In slightly less than forty-eight hours, of which he had slept for twenty-seven, the little bear had come to accept us as his protectors, reacting now much as he would have done with his mother upon feeling in need of security. I knew from past experience that extremely young wild animals settled down quickly when fed and comforted, but that older animals of an age comparable to that of the little bear's usually required four or five days in which to develop their trust. Young raccoons at approximately the same stage of development as Snuffles had hitherto taken at least four days to accept us fully, in some cases as long as a week. Our latest ward, it seemed, was unduly quick to grasp the fundamentals of survival!

When Joan signaled that all was ready, I carried the cub into the bathroom, disentangled him gently from my body, and lowered him into the warm water. Lulled by his tranquillity and the trust he had demonstrated, I became careless, failing to hold him securely as he encountered the liquid. The speed of his reaction caught me completely off guard. Like a reluctant swimmer who tests the water temperature with one toe, then escapes to a warm place on the sand, Snuffles barely got his back feet wet before he squirmed out of my grasp even as he wrapped himself around my right arm, from there to scurry swiftly upward until he found refuge on the back of my neck, his back paws digging for purchase against my vertebrae and his arms wrapped tightly under my chin.

Joan was amused once more. I wasn't. But I couldn't get at my unwelcome passenger in order to unwrap him without drawing blood. I asked my wife to help—twice. When she ignored my pleas—excusing her lapse later by claiming that she hadn't heard me above her laughter—I turned, aiming my messy passenger in her direction, and backed toward her. Caught between me and the doorway and only about three feet from the toilet, she was unable to evade our ward when, on finding himself near her, he forthwith transferred his allegiance, releasing my neck, whipping around, and drop-kicking himself away from me as he grabbed her coat, up which he scurried until he had attained his favored position, head pressed against her chin and neck. Joan stopped laughing now. Next she commanded me to "get the little swine" off her. This time she meant the insult.

Smiling broadly and telling her that she should not spurn the bear's love with harsh words, I unwrapped the cub and secured a good, firm grip on him before putting him back into the water. He fought, of course, as did all those wildlings subjected to the same treatment, but he had to submit while a none-too-gentle Joan applied shampoo and began to lather him. When the half that projected from the water resembled a mound of whipped cream, I lifted him out and put him on top of the board that Joan had placed across the tub, a platform upon which we anchored unwilling bathers while working shampoo into their nether regions. Snuffles was no longer placid; he even screamed, a cry of anger rather than one of anguish. Then he bit Joan on her thumb, quite deeply.

Matters became somewhat confused as I secured a fresh hold on the bear's slippery neck fur with one hand and let go with the other so as to allow my wounded wife to pour shampoo into my cupped palm. While this was going on, Snuffles danced and wailed in rage, but oncc I could work with both hands, rubbing the soap into him and scratching gently deep down into his underfur, he calmed to some extent, though he did not give up his struggle for freedom.

Snuffles finally emerged from his bath fluffy, sweet smelling, and not any the worse for the experience, the last part, which included toweling and a brushing with my personal hairbrush, returning him to good humor. While Joan, her thumb patched with a Band-Aid, cuddled our bear on her lap in the living room, I cleaned out our bathtub, first scooping out the two floating islands of dung that Snuffles had left behind, then scouring, disinfecting, and rinsing the entire enameled area.

Cleaning up the kitchen was not such a simple operation. It was hindered by the one responsible for the mess, who, being now clean, dry, well rested, and replete with good things, wanted nothing more than to continue to explore his newfound environment. He could have been put in the barn while we both tackled the destroyed kitchen, but we were afraid that he might find his way out of the building by climbing into the hayloft and emerging through its loosely fitting doors.

I knew now that I had badly underestimated the cub's recuperative powers! The night before, I had made plans to construct inside the barn a temporary cage for Snuffles by cordoning off with heavy-gauge chicken wire an area ten feet square and six feet high, within which I would place a leaning log or two in addition to some other natural playthings. I had proposed to tackle the job right after breakfast on this day.

As matters turned out, Joan held our ward on her lap while I slaved in the kitchen, scraping up the mess, mopping the floor, walls, and all other objects therein, and almost having to completely dismantle the stove top before all the goop, glass, and lumps were removed. When the room was, if not pristine, at least habitable, I took a turn at holding Snuffles while Joan cooked bacon and eggs and made coffee and toast. We breakfasted late, all three of us.

By the time I got around to feeding the rest of our menagerie that day, our natives were restless. The majority of these, fortunately, were free agents who no longer needed continuous care, but there remained two lodgers who, despite being able to garner their own wages, insisted on being subsidized. One of these was a yearling red-tailed hawk; the other was a yearling female skunk.

Of course, one never knew from day to day, or even from moment to moment, when one of our independent, part-time residents would show up seeking a handout. So, like a Boy Scout, I was prepared each morning when I emerged from the house carrying feed. In any event, we catered to numerous and assorted birds and to an aging chipmunk named Scruffy, all of whom were what might be called permanent guests.

Since they demanded it noisily, I fed the birds first, as usual escorted by our resident flock of chickadees and accompanied by the screams of the jays. The chickadees, because they are such trusting little birds, landed on my head and shoulders, the highest-ranking members of the pecking order flitting down to help themselves from the seed pail and even from the scoop as it was lifted from container to feeder. The nuthatches, not quite so daring, sat in the trees nearby, filling the midmorning with their harsh *yank-yank* calls, and the flock of big jays flitted back and forth impatiently, rowdy creatures who would not still their raucous yells until each had crammed itself full.

Leaving the empty seed pail beside the porch, I carried a second one to the barn area, where I expected to meet Penny, the skunk. Usually, she lived under the machine shed and would emerge during daytime when she heard or scented me. Of late she had been otherwise occupied giving birth to, and caring for, her first family, though the exact whereabouts of the nursery was unknown to me. For maternal reasons, Penny's timing had been off during the last week; she was now liable to show up at any hour between seven and noon, gulp down the mixture of egg and ground beef prepared for her, then waddle away toward the grove of young maple trees in the shelter of which she had delivered her babies. As a result, I now limited myself to putting out her food and collecting the tin plate later in the day, suspecting that the jays and other hungry ones might well steal most, if not all, of it.

Penny had come into our care the previous June, toward the end of the month, when she was about seven or eight weeks old, a diminutive, undernourished little striped skunk whose mother and siblings had been run over by an automobile, their highly odorous carcasses spread all over the road. A teenage boy, son of some distant neighbors, had come by on his bicycle and jumped off the machine so as to wheel it and himself into the roadside forest in order to avoid the smell of skunk that filled the air. While engaged in this detour, he found the small skunk huddled against a tree trunk, and because she didn't raise her tail and made no attempt to do so when he approached and, finally, touched her, he picked her up. Penny remained docile. Billy was carrying a satchel affixed to his bicycle's saddle; he put the skunk in it and rode for North Star Farm, delivering the sad little creature to our care.

The way Billy explained it to me, it sounded as though the mother and four members of her family had only just been killed; but Penny's physical condition was such that I doubted this; so I climbed into the car and drove over to have a look for myself—to determine whether the young skunk's condition was due to starvation and neglect, or to disease. If the former, it meant that her mother had been dead for several days; if the latter, the skunk would have to be destroyed, for these animals, apart from being subject to rabies, are also victims of a variety of other pathogenic agents the spread of which would be disastrous to the other inhabitants of our farm sanctuary. At the scene of the slaughter, I needed but one look to determine that the animals had been killed several days earlier. The young skunk was weak from hunger and probably suffering from shock.

Penny's life hung in the balance for six days, but she slowly responded to nourishment and kind treatment and eventually became one of the most charming and docile inhabitants of our ménage. As the future was to show, she never sprayed anybody unless she was given very good reason for so doing and thus was always welcome around, and inside, the house. While she was convalescing, the only times that she became slightly odorous was when constipated, on which occasions the need to strain squeezed some of her powerful musk from the twin glands concealed within the rectal region. Once I had found the right dietary balance for her, this problem vanished and she remained as sweet as a nut, almost. Penny was christened by me one morning, about two weeks after she had arrived, while I was pondering her virtual lack of odor.

"She only gives out a little scent," I told Joan at the time. That statement suggested her name. Joan liked the name, but, as often as not, insisted on calling the skunk Penelope, claiming that the way in which I arrived at her name was "so devious, only *you* could think of it."

However that may be, Penny, once recovered, progressed in leaps and bounds, literally. She hunted like a terrier, pouncing on anything that moved through the grass if it was small enough; she caught mice, beetles, moths, grubs, and so forth. As a result she was a welcome visitor to Joan's flower and vegetable gardens, or what was left of them when the groundhogs had finished eating. In addition to this side benefit, Penny was also my secret weapon for those times when uninvited and tiresome visitors continued to overstay a welcome they had never really received. Despite assurances to the contrary, such visitors felt immediately apprehensive when they learned that Penny had not been de-scented. When the skunk was brought into the living room and introduced, it took them no time at all to remember that they had other appointments.

In fact, Penny, like all of her kind, was a docile and gentle soul who liked to mind her own business as she waddled placidly through a life that consisted mainly of eating, sleeping, and eating. Now and then, when one of our other wards became impertinent with her, she would teach them a swift and stinky lesson, but even at these times she showed restraint, giving them a short "puff" from only one of her versatile guns.

Skunk "shooters" are operated by muscles that surround the gland and allow the animal to fire a twin salvo, operate on a single shot, and alter the angle of fire. They can squirt out of each muzzle directly at two different targets, or they can turn each jet into a choking spray by aiming so that both streams collide a few feet away from the gunner; this collision creates the vapor. Effective armament indeed! Even so, skunks are vulnerable to many predators, including lynx and bobcats, cougars, wolves, coyotes, and dogs. In the case of wild predators, hunger usually determines an attack; if the hunter is really hungry, it will risk the musk bath and will even eat the shooter, spray and all; where dogs are concerned, some of these are so carried away with an attack that they kill the skunk before becoming as "sick as a dog."

Mature female skunks in our latitude usually mate during February, but young females come into their first heat in March, at the tender age of ten or eleven months. Penny evidently succumbed to the charms of some local skunk Lothario in early spring and of recent days had become very bulgy as she approached the end of her eight-week confinement. About a week before Snuffles came into our care, the skunk failed to show up for three days; when she did eventually appear, she had recovered most of her waistline, but she now carried six swollen, milk-tipped little dugs.

Snuffles, the bear, relaxes in a tree.

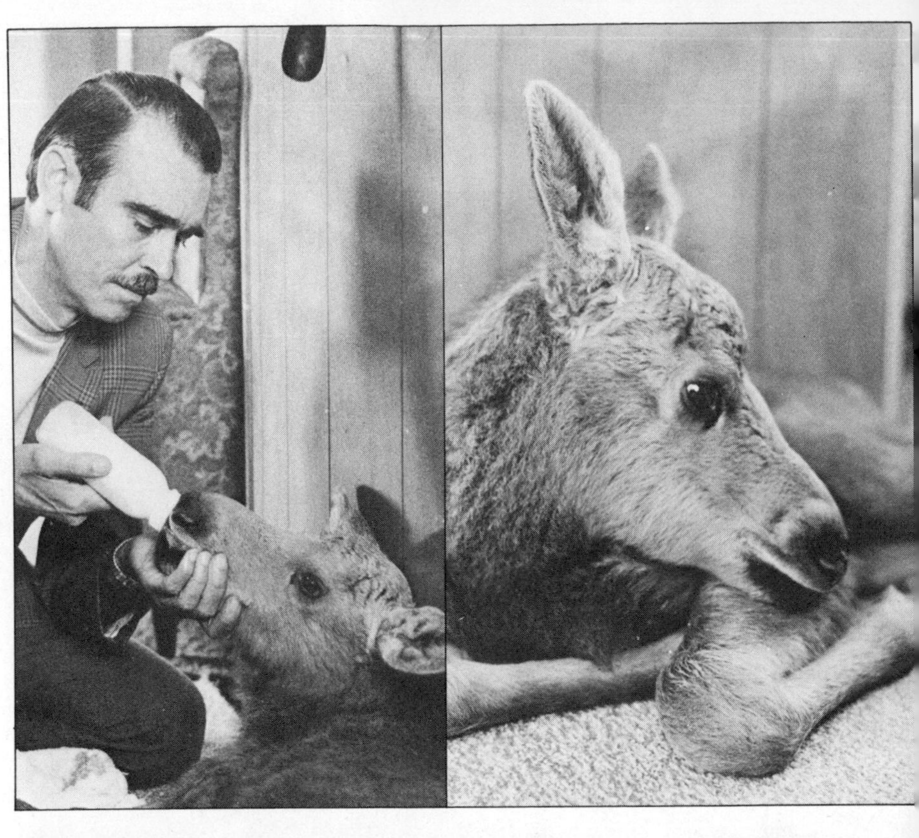

Babe, having a good scratch.

FACING PAGE:

Top left: *Snout, the moose, accepting his first feed from the author soon after being rescued.*

Top right: *Snout, one month after arrival at the author's farm, relaxing in the kitchen.*

Below: *Snout, fully adult, feeding on water plants in the deep wilderness during summer.*

Slip, left, and Slide, river otters who refused to be domestic pets.

FACING PAGE
Above: *Colby, top, and Conk at breakfast.*

Bottom left: *Herself, visiting the farm kitchen at night.*

Bottom right: *One-Ear, the male raccoon, who lost his ear during an altercation with Snuffles the bear.*

On Day Three Post Peace, Penny was waiting for me at her accustomed place, her silky, black-and-white tail held upright, its tip curving downward, the "happily-at-ease" position. Before I had stooped to take her food out of the pail, she was rubbing against my legs like a cat and, when the plate was before her, even as she munched, she arched her back to my gentle scratching. Moments later I left her, angling toward the maples, where I was sure her young now lived, to carry the last rations of the morning, some chunks of stewing beef for the hawk.

Maggot was about the same age as Penny, but whereas she had left us in the fall to find her own winter quarters, the hawk had been kept inside a large, aging henhouse, there to bask away the winter under the benign rays of a heat lamp, while his less fortunate relatives undertook the long and strenuous migratory flight to the southlands.

Maggot was brought to me the previous summer by a neighbor. The young red-tail hawk apparently had fallen out of his nest during flight training and had injured his right wing, cutting it quite badly on some of the sharp, dead branches through which he had dropped. The bones were uninjured, but the wounds were infected; as a result he was unable to fly south when the time came. Now, after spending the entire winter in the lap of luxury, Maggot was not inclined to feed himself, preferring instead to come to the house area, perch in one of several favorite trees, and whistle his high-pitched, rasping call—*keee-errr*—until I brought him his meat.

To encourage his independence, I recently had placed the hawk on short rations, and this, judging from the debris that now reposed at the base of his favorite roosting tree, appeared to be having some effect. Woodland hawks, or buteos as they are properly called, rarely settle to eat on the forest floor, but carry their food into one of several "plucking" trees, which generally are selected because they are dead and leafless and thus provide a good view of the surrounding wilderness. On top of such perches, the hawks pluck off feathers and fur with their beaks as they consume their food, the leavings of each meal falling to the ground. In Maggot's case, these leavings mostly were made up of mouse fur and bones with occasional fragments of chipmunk hide and fur. He was not yet doing very well, but he was learning; soon, I hoped, he might go his own way—especially if some unattached female red-tail happened to fly into our area. Meanwhile, I fed him and was always amused when he would land within a couple of feet of the "prey," stride forward, wings akimbo, and then strike with one clawed foot, killing the quite dead stewing beef chunk by chunk. Never, since he was old enough to feed himself, did he fail to "strike" and "kill" before grasping the meat with his beak.

Maggot was a most interesting bird. I suppose I tended to spoil him because of this and also because he taught me something about nature that had hitherto been concealed from me—and I suspect from many other biologists even to this day.

He came to us in late afternoon, a wild-looking youngster, almost fully fledged, with the glaring eyes of his kind and a tendency to clench and unclench his formidable talons in a manner that made his handler distinctly nervous. To treat his wounds, I first tied his legs together with strips of soft cloth, then pulled a sock over his head, the first to immobilize his inch-and-a-half *very* sharp claws, the second to keep him still, for when a bird cannot see, it usually remains docile.

After cleaning and disinfecting his injuries and noting that no bones were broken, I untied Maggot and settled him in the henhouse, where a number of wrist-thick branches offered a variety of roosting places. When I felt that he had been allowed enough time in which to relax, I returned with cubes of stewing beef, taking the first piece out of the dish and profering it to him with my fingers. Turning his head sideways, the better to fix one glaring orb on the food,* the young hawk returned to a face-on stance, leaned forward, and stabbed with his curved beak, impaling the meat as neatly as a charging lancer spears an enemy, mercifully missing my fingers! Twisting sideways, the hawk transferred the morsel from beak to talons, then dissected it bit by bit and with surgical precision, swallowing between each peck. I had thought it would have been necessary to coax him to eat, perhaps even to force-feed him, but Maggot was evidently extremely hungry, or very aggressive, or both. Yet he was also careful to take only the beef, leaving my fingers unscathed.

The next day, when I went to visit the patient, it was to find him trying out his wings, the good one flapping well, the injured one only partly rising, causing the bird to lose his balance and causing *me* to reach for him before he crashed to the cement floor. We caught each other at exactly the same time. As my hands closed over his body, one of his rapier-equipped toes stabbed my right thumb, piercing the skin and going through it like a suture needle. In this way we were not only united, we also became blood brothers.

Getting the talon out without exposing my left hand to Maggot's other foot was a thrilling endeavor that ended successfully when I decided that finesse could not be employed. On the premise that either his leg or my flesh *had* to give, I gave a tug. His leg did not give.

*Birds do not have binocular vision; each eye works independently of the other. For this reason, they cannot effectively focus on a near object. This is why birds peer "one-eyed" when looking at something that is close to them.

When the blood was staunched by a Band-Aid and the hawk was once more settled on his perch, I prepared a meal for my erstwhile attacker and took it to him, accompanied by Joan who came, she assured me firmly, *purely* as a spectator. Maggot eyed us both fixedly but remained placid. Again, he quickly consumed the food that was held out to him.

Watching him after he had ingested his full rations, I noticed that a number of maggots were crawling through his feathers, most of these congregating on his neck and head. Joan noticed them also; she backed closer to the door and eyed our latest guest with considerable distaste. Maggot, meanwhile, turned his back on us, his action allowing me a better view of two particularly fat "worms" that were frolicking among his head feathers. Concentrating as I was on the grayish wrigglers, I failed to notice that the hawk was going into what I call the "toilet crouch" adopted by young birds of prey for extremely hygienic reasons. Joan later accused me of owning a perverted sense of humor, claiming that I knew perfectly well what the "filthy bird" was going to do. Had I noticed the hawk's crouch, I would have known, but being preoccupied with the maggots, I honestly did not take cognizance of the bird's intentions.

Indeed, my first awareness came when a liquid, gassy sound exploded not far from my face only instants after Maggot had elevated his posterior and fanned his tail. Following the warning sounds, a stream of chalky, runny stuff was forcefully ejected by powerful, special muscles in the bird's unmentionable part. The volley was fired with deadly aim. It was propelled swiftly through eight feet of space and hit my wife in the chest with audible impact. For a time after that I forgot about the maggots.

Later, remembering the intriguing worms, I returned to the hawk's quarters, alone this time, with a large magnifying glass. The bird sat staring into space, turning his head jerkily now and then, but at peace with the world. I approached him, lifted the glass, and began studying his passengers. One maggot fell off. It landed on the perch, a three-inch, bark-covered branch, where it stuck. Maggot turned his head, peered one-eyed at the wriggling thing, and then calmly lowered his beak and ingested his recent lodger. Fascinated, I returned to my observation of the fly larva.

To my astonishment, it at first appeared as though the maggots were eating the ends of the hawk's feathers; but after longer and closer observation, I at last understood. The rather revolting larva were *cleaning* the hawk's plumage by eating the carrion that had adhered to them!

Perhaps others are also aware of this interesting action of an otherwise unprepossessing insect, but I had not been aware of it at the time, nor have I since read about it in the literature. Yet it is all so logical when one thinks about it. Young hawks spend a number of weeks in untidy, stick-built nests that offer ideal hideaways for fallen bits of meat. The eyases (as young hawks are called) slowly learn to pluck fur and feathers off their food as they develop skill with their beaks. During this "apprenticeship," a fair amount of waste ensues and much of it lodges in the interstices of the nest, where it decays. Then come the flies, the females of which retain the fertilized eggs inside their bodies until the maggots are hatched; the maggots are laid on rotting matter, cleaning up the leftovers and even crawling into the feathers of the young hawks, from which they remove bits of putrid meat, blood, and other vital juices that have now become caked. Thus the maggots are, in effect, housekeepers for the hawks; but their role doesn't end there. Some of the larva undoubtedly complete their development inside the nest, but many are eaten by the eyases, offering them additional food that is rich in protein.

Not wishing to draw final conclusions from that one experience, I later developed the habit of examining hawk young and their nests, inspecting at close quarters sixteen nests of three species of buteo: broad-winged (two nests); red-tailed (six nests); and red-shouldered (eight nests). In all, over a period of five years, I examined forty-six young hawks in stages of development ranging from down cover to nearly fully fledged. In every instance maggots were present in the nests and on the young, the insect larva behaving in much the same manner as did those first noticed on Maggot's body. Since my initial supposition was now confirmed and because the scaling of tall deciduous trees in order to study the dwellings of woodland hawks is an arduous pastime, I abandoned the practice.

In numerous and ingeniously effective ways, Nature causes her creatures to keep a tidy house; the maggots that eat the carrion from a hawk's feathers, and from inside its home, represent one of these. Another stems from the manner in which young hawks (and other large birds whose chicks are unable to fly for several weeks after hatching) are capable of expelling their feces over the rim of their nest rather than allowing this to fall into the cup, wherein the runny, limey discharge would cake their bodies and destroy the insulating and waterproofing qualities of their down or their feathers.

Such birds are equipped with especially powerful rectal muscles and an evidently inherent need to raise their behinds and fan their tails so as not to soil them when they are gripped by the urge to defecate. The "tip-up" movement aims the derrière, while the sudden spasm of the muscles expels the droppings with more than enough force to clear the nest rim and to sail some distance beyond it, as my wife discovered when she happened to stand in the wrong place. But to get back to the maggots . . .

In their honor did I name the hawk, though at first I considered adopting a variation of the fly's family name *Sarcophagidae,* which I translated as Sarcophagus (Gus for short), but on mature reflection, it seemed that Maggot suited him better. Joan, however, had another name for him, but in view of the vendetta that developed between my wife and my bird, I considered her choice biased, although it cannot be denied that Maggot, deliberately (as Joan claimed), or otherwise, turned his back on her when she came into his presence—which she did only in my absence and then for just long enough to allow her to *throw* his meat inside the henhouse—and immediately thereafter adopted the toilet crouch. But after that one experience, Joan was always too quick for Maggot, slamming the door shut before the bird fired his shot. For purely scientific reasons, I wanted to paint a bull's-eye on the door, to see if Maggot was a good marksman, but Joan said that if I did that, she would never so much as open his door.

"And he can *starve* to death when you're not here, for all I care. He's *revolting.* And so are you!" she concluded.

On the particular day that Snuffles wrecked our kitchen and bit Joan's thumb, Maggot was sitting high in a maple, screaming insults at me while he watched my progress across the open ground. When I stopped on the edge of the trees, he launched himself into the air, planed downward, and landed with wings outspread, waiting for his delayed breakfast. I put down his tin plate of beef cubes and left him, anxious to get going with the construction of the bear's cage.

Three weeks after the bear's arrival we had to make a journey necessitated by the fact that one month before rescuing Snuffles I had visited a malamute breeder and picked out a tiny male puppy who was intended to be a companion for Joan, a protector in my absence, and a transporter of my camera equipment (in saddlebags) on those occasions when I would make a foot foray into the wilderness surrounding my farm intent on setting up a blind for photographing some animal. As matters turned out, he admirably fulfilled the first two requirements, but failed miserably in carrying out the third, perhaps because he felt it was beneath his dignity to shoulder a burden, or perhaps because during his infant days at the kennel he had developed a fondness for canvas and leather; however it was, he either charged away after a mouse or some such worthy foe, or else amused himself by eating his backpack while I was busy constructing the blind. After the second time, it was more expedient to carry my own loads and leave Tundra his role of wife protector, which was something that he was good at. But that was later, after he had grown up. Meanwhile, the time had come to go and collect him, for he was nine weeks old and ready for his life of adventure at North Star Farm.

At first I intended to go alone, leaving Joan to care for the farm and, more particularly, for Snuffles, but by this point in their relationship, Joan and the bear, though extremely fond of one another, tended to engage in arguments that resulted when Snuffles wished to sample some delectable thing that was forbidden him; such altercations invariably took place during my absence. Just as invariably, they ended after either Snuffles had had his way or our home had suffered considerable upset.

Let there be no mistake about it, Joan loved Snuffles! But the rambunctious and very healthy cub respected only one form of discipline, the kind that a mother bear reserves for those occasions when her offspring fail to behave: a good, hard swat with a big paw that usually sends the offender tumbling head over heels, screaming blue murder. Joan was just too gentle for such rough stuff; instead she sought to reason with the offender, who went along on his merry way; as a last resort she would seek to drag him away and to put him outside, whereupon Snuffles forgot his appetitive desires and devoted himself wholeheartedly to the game he thought Joan was initiating. First he wanted to wrestle, and if, as so often seemed to happen, my wife's legs were clad in nylons, these delicate stockings were reduced to shreds while the skin of calves and thighs became crisscrossed with scarlet lines. When such frivolity was discouraged with a few pats of a broom, Snuffles would try another gambit; he would race through the house, over the furniture, and even under the bed.

After several demonstrations, Joan refused to bear-sit and insisted that the orphan be held in jail until I returned from wherever I was going. This worked well, until the night before I was due to collect the pup, when Snuffles amused himself by eating the chicken wire out of one corner of his prison and discovered early next morning that he was *really*, *really* fond of potted geraniums. Since I had no time to fix the hole in his cage, and inasmuch as Joan refused to be left alone with him, we all had to go. This seemed like a simple solution to our problem.

Because of the impedimenta of my outdoors occupation, I stick to station wagons for transportation purposes, so the prospect of taking one small bear cub on a 250-mile round trip in such a vehicle did not pose problems in logistics. Even allowing for both bear and dog, plus food for humans as well as animals, there was, I thought as we started out of the farmyard, plenty of room for all.

Snuffles, busy examining the interior of the station wagon, failed to realize that the vehicle was moving at first. When he did, he seemed to enjoy the experience and bounced his way to the back seat, on which he perched while watching the moving scenery. At this stage, Joan was sitting beside me, but turned to face the back, keeping an eye on the bear while I, sparing him occasional glances in the rear-view mirror, concentrated on driving.

For almost an hour, while we drove over country roads, Snuffles was as good as gold, alternately gazing out of the windows and playing on the floor of the wagon with a large rubber ball I had bought for him ten days earlier. We were beginning to feel quite proud of our ursine foster child.

Eventually the quiet roads gave way to a busy, six-lane highway, the main east-west route across the southern part of Ontario. Lulled into a false sense of tranquillity, we spared Snuffles hardly a thought as I drove down the entry ramp, gunned the motor so as to get into a lane before some insane motorist rammed us in the tailgate, and then settled the car into a safe and comfortable fifty-five miles per hour. And then a transport truck came thundering by . . .

It was hot, so the car windows were open a third of the way all around, including the tailgate glass. I was paying strict attention to the business of driving on the freeway, relying on Joan to watch Snuffles; but my wife, herself made nervous by the roaring truck that approached, drew level, and eventually passed us, was not watching the bear. As usual, the great machine's air displacement rocked the station wagon, but this would not have created a problem under normal circumstances because I was expecting it. Indeed, I was engaged in correcting the steering when I was suddenly blinded by two clutching, hairy arms. The car dived to the right, riding the grassy verge that sloped toward a deep ditch; I lifted my foot from the accelerator and started to apply the brakes; at the same time I tried to control the wild steering wheel with one hand while the other attempted to push up the blindfold. My mouth, however, was not impeded; I yelled at my wife, uttering a terse, four-word sentence: "*Get him off me!*"

Joan did as she was told. I nearly lost an eye when one of the "blindfold's" claws scraped across my right lid as it traveled upward, then hooked temporarily in my eyebrow, opening one corner with the precision of a surgeon's scalpel.

Mercifully, the car had slowed almost to a stop. I halted its movement altogether, because now, though the cub no longer gripped me tightly around the head, the blood from my cut brow was effectively blinding my right eye. I was *not* in a good mood. I was as angry with my wife as I was with Snuffles, both of whom I addressed in language learned during long years in the army. Alas that such eloquence should be wasted on its intended audience! Joan was far too occupied with Snuffles to pay any attention to my words; Snuffles was too terrified to pay attention to anything. All his energies and attention, I noted out of one eye as I glanced over my shoulder, were being expended on the task of crawling up my wife's back while he discharged a goodly amount of fear-induced diarrhea all over everything, including Joan's blouse. It was a scene right out of Bedlam!

The next twenty minutes were spent putting things to rights, parked on the verge. Joan had sensibly brought a complete change of clothing in addition to her "critter" smock, putting on the latter after she had mopped up the Snuffles mess, sprayed the interior of the car with disinfectant, and lowered the back seat to enlarge the loading platform. Meanwhile I attended to my eyebrow. It wasn't much of a cut, but, like all head injuries, it bled profusely; peroxide and cotton batting from the first-aid kit cleaned up the mess, and some styptic checked the bleeding immediately.

During the fracas, Snuffles managed to get himself well smeared with his own wastes and he needed mopping up. I devoted myself to this task while Joan prepared half a bottle of formula for him in the hopes that he would go to sleep for the remainder of the journey. While the cub was feeding, we discussed tactics. Henceforth, Joan would ride in the back, sitting on the platform and leaning against the front seat, from where she could keep an eye on Snuffles and prevent him from repeating his headlock on me as well as offer him comfort when noisy vehicles overtook us.

The remainder of our journey to the kennels was only relatively uneventful. Snuffles continued to react fearfully when the traffic was busy, but he soon learned to rush to Joan, who hugged him to her and covered his head, and especially his ears, with her hands.

Greeted by the prolonged howling of thirty-odd Alaskan malamutes, we drew up outside the kennel office and were met by the breeder who, stout fellow, didn't so much as raise an eyebrow when he saw our small passenger. Fifteen minutes later a bouncy, inquisitive puppy was introduced to a bouncy, inquisitive bear cub, the meeting accompanied by much canine tail wagging and a great deal of ursine snuffling and nose prodding. The two accepted each other instantly and were engaged in play before we cleared the environs of the kennel. Joan sat in her place, smiling benignly; I felt pretty good, for now that Snuffles had a companion, I was sure that he would forget his earlier fears. And so he did—until we reached the confounded freeway and the traffic noise activated him anew. On this occasion matters went from bad to worse because the pup also became afraid.

Each reacted almost identically by first defecating, then rushing to one another, the dog trying to crawl under the bear and the latter seeking to hide himself under the dog while both of them bounced over the loading deck. Joan alternated between trying to mop up the messes and seeking to comfort the terror-struck duo, but after such performances were repeated three times, each sparked by big, roaring trucks, I concluded that unless we got off the freeway and picked our way back along country roads, our chances of ever seeing our farm again were slim.

We added about fifty miles to our journey by taking the detour, but our passengers settled down considerably. That is to say, they showed no fear. Instead they engaged in play, and if this caused Joan to intervene frequently, when one or the other got roughed up and began to squeal, her involvement did not include the cleaning up of solid wastes. It *did* include the mopping up of liquid waste. When we were still some sixty-five miles from home, Joan opined that she had had enough. She would drive; I could ride in the back. We were about to switch when we came to a small city; this delayed the trade.

Entering the built-up area, we were halted by a traffic policeman who was doing duty for the signal lights, which were out of order. Traffic was not heavy; the policeman was clearly bored and taking his time as he directed the vehicles. Evidently preparing to wave us on our way, the minion of the law stepped backward, placing himself about three feet from the right-side window of our car. This seemed to interest Snuffles, who lumbered over, climbed on the back of the driving seat, and stuffed his nose through the partly open glass before cutting loose with one of his calflike calls. As swiftly as the cub climbed up, he slid down, this time landing on the floor, passenger side, next to my legs. The policeman turned and saw the pup, who had risen on his hind legs and was staring through the glass. He grinned, then waved me on, but as the car moved forward he bent down and called out, "That's some bark your pup's got!"

A little while later we stopped, switched drivers, and continued a journey the like of which I had never undertaken before, have not experienced since, and which I fervently hope I never shall undertake again. By the time we arrived home our vehicle had acquired a mephitic fragrance that was to take weeks to eradicate and the authors of the effluvium were in desperate need of shampoo and lots of water. As for the human protagonists, we were exhausted and definitely neurasthenic.

It may here be noted that Snuffles never again set foot inside our automobile. The dog, after he grew some, learned to behave inside a car and positively delighted in doing so. Subsequent to that ill-fated journey, I have carried a number of other wild ones in my vehicles, including a two-hundred-pound lion and, once, three young chimpanzees: none of these creatures created as much havoc as Snuffles and his pup companion.

Four

The lynx sat inscrutable on the porch roof; still as a sculpture, he watched as the station wagon came to a stop, and as we got out of it, leaving our two grubby passengers shut inside for the time being. Neither of us noticed the big cat at first, but as I walked toward our home some extra sense told me that I was being intently scrutinized. I looked up. His enigmatic, amber eyes fixed themselves on mine and followed my progress; beyond that, he didn't move as much as a whisker.

It was late afternoon; the sun was low behind the house, casting a deep, rectangular shadow in which the lynx blended perfectly, his pastel fur, with its shadings of gray and its streaks of black, combined with his utter immobility to lend him a ghostlike air. But there was nothing phantasmal about his penetrating orbs; attracting and returning the ebbing daylight, the glowing eyes gave substance to the being that waited with inherited patience for the food that it knew would be forthcoming now that we had returned home.

Just before entering the porch, I stopped to speak to our onetime guest, my words meaningless to him, yet my voice causing him to respond by lifting his haunches and adopting an upright posture, his movements soundless, but deliberate; and he purred, a loud and deep rumble so similar to, and yet so unlike, the sound produced by the domestic feline. As I stepped into the porch, key in hand, I heard the whisper of his great pads as he glided across the lean-to roof preparatory to his leap to the ground. A moment later Joan spoke to him also, asking where he had been "all this time," for we had not been visited by Manx since early spring.

We estimated that the lynx was now almost four years old. He had been earning his own living for the last two years, ever since he became fit enough to survive in the wild, after coming to us contained in a stout, slatted, wooden cage, snarling and spitting like evil incarnate, a yearling, but old enough to make his own way in the forest—except that he was injured.

Manx was one of a number of lynx that had been trapped in the north of Ontario and brought to a region some miles distant from our farm, there to be released as part of a program aimed at reestablishing the breed in an area where they had once lived, but from which they had been exterminated by trapping and hunting. Officially, the young lynx had been set at liberty with the other captives; unofficially and, in fact, because he had cut his thigh and wrenched the tendons of his left leg while contained in the trap, he was brought to me by a conservation officer, a government employee who was too humane to condemn the injured animal to a slow death in the wild merely because regulations dictated that he should be released. We agreed that Manx was to be given his freedom as soon as he was able to hunt for himself, and that I would not write about the lynx in my newspaper column because this would compromise the cat's benefactor, promises that were kept; but the second one no longer binds me, for my friend has retired from government service.

Young as he was, Manx was big enough to be dangerous, so before he could be removed from his cage it was necessary to find suitable quarters for him. Once again the old henhouse was pressed into service. The building was made of hand-hewn logs, great timbers of seasoned maple that the land had supplied for the original homesteader some one hundred years earlier. It had a solid door fashioned out of home-sawn, two-inch planks and a square window that, long before the lynx arrived, I had covered with heavy-gauge wire mesh. In addition, its builder had incorporated a small hatch on the south wall that allowed the hens to enter and leave at will; this I had adapted to suit the less orthodox livestock that the building housed at intervals, enlarging the entrance, making it twelve inches wide by sixteen inches high and fitting it with a sliding steel door.

While Joan sought to calm the thoroughly aroused cat by talking to it quietly and continuously, squatting about two feet from the cage, I remodeled the interior of the henhouse, removing the perches that had been put there for its last occupant, an ailing great horned owl, sweeping the cement floor, cleaning out the water container that could be filled from the outside, and hauling into the building a number of logs; several of these were spiked into the walls so as to furnish off-the-ground vantage for the cat while the remainder, which were shorter, were placed at an angle in one corner, making a dark sanctuary that was then filled with straw. The last thing to do before transferring our new ward to his temporary quarters was to get from the freezer an entire beaver haunch. This was frozen, of course, but if the cat was hungry he would eat it even in that condition; in any event, it would certainly be thawed by early morning.

Joan's efforts to calm the lynx had been unavailing. He continued to snarl and spit, alternating these ill-mannered noises with rumbling growls that seemed to issue from his stomach rather than from his throat, his tufted ears sloping backward, his stiff whiskers pointing forward, and his great mouth agape and filled with wicked teeth.

To prevent his snacking on one of my legs, I tried to introduce him into the henhouse through the outside hatch, placing his cage against the opening, then raising the sliding door, but he backed against the far slats of his prison, refusing to emerge as he increased the tempo and ferocity of his snarls and growls. When after several minutes it was obvious that he was not about to change quarters by present means, I was forced to carry his cage inside the henhouse.

Shutting the door, I stood behind the cage and again pulled up the slide that contained him. The lynx didn't cower away from the exit this time; but he didn't leap out as I had expected him to do. Instead he moved forward cautiously, stuck his head through the opening, paused briefly to glance around, then bounded forward, his movements awkward because of his injured leg. A moment later he disappeared into the den that I had made for him. I had anticipated problems, even feared that he would attack me, but it had all been so simple!

Outside, Joan whispered through the door.

"Are you all right?"

Her voice told me that she was "palpitating," as I referred to her rare moments of stress. It also told me that she too had considered the possibility of my being attacked by the seemingly vicious lynx. Because I didn't want to disturb him now that he was settled and quiet, I started to open the door instead of replying to her. The hinges squeaked, so I pushed slowly, trying to be as quiet as possible. When my wife caught sight of me I knew by the roundness of her eyes that she had imagined me instantly consumed by the lynx and that it was the cat, and not her husband, that was about to emerge through the entrance! Allowing her to scold me a little so as to relieve her anxieties, I secured the door and we left the irate cat to its own devices.

Later, seeking a name for him, we settled on Manx, in honor of a breed of bobtailed cats native to the Isle of Man that lies off the northwest coast of England; the people of this island speak a language all their own, known as Manx. Likewise, the unusual native felines are called Manx cats.

Manx was not the best patient we ever had. Indeed, for the first six days, we never saw him, his continued presence being confirmed only by the disappearance of the meat that I put into his dwelling each evening and, later, by the distinctly rank aroma of the raw leftovers and of his toilet arrangements. Somehow, it was going to be necessary to win his trust, and that soon, for we were worried about his injuries as well as by the need to clean his home. As a last resort, I could mix a tranquilizer with his food, but I was loath to do this because only a dose capable of knocking him out would serve the purpose and such a quantity could be dangerous.

After much debate, it was decided that I would not feed him for two days, then I would enter his house in early morning, carrying a good supply of prechopped beaver meat and would clean the place and sit there, the food between my legs, and wait until hunger brought him out of hiding, at which time I would feed him piece by piece. This stratagem worked, although it took three days before Manx allowed himself to approach near enough to me so that he could take meat from between my fingers. After that, patience combined with continuous hand feeding slowly won him over.

Ironically, by the time the day came when he allowed us to touch him, showing pleasure in our stroking and scratching, Manx's wound had healed of its own accord. But he continued to limp for almost a year, during which time he slowly widened his explorations of the area around the farm while returning each evening to get his food. Eventually, he chose a territory for himself and only came to visit us on rare occasions; but he retained his trust, though after he had been living on his own for about six months he no longer allowed us to touch him. Instead he purred, telling us that he would be obliged if we would ask him to stay for dinner.

We were not totally overjoyed to find the lynx waiting for us on our return from the kennel. We were glad to see him, of course, but we knew from experience that he was likely to hang around the farm for several days, prowling mostly between dusk and dawn, but often enough turning up during full daylight. Now that we had to care for the bear cub and the puppy, we were worried in case one or the other of them met the lynx, who was not the sort of animal to be trusted with such young and inquisitive beings. The cat probably weighed forty pounds; his armament was as formidable as his temper was short. It was not likely that he would be able to fatally injure Snuffles, but, given an opportunity, he might well kill and take away the pup, to eat him in the shelter of the forest. Until Manx volunteered to return to his own territory, we were going to have to keep a close watch on our two "babies."

On the premise that if the lynx was given an extra large feed today he might be satisfied to return to his home range during the night, I took him more food than he would normally consume. He was waiting for me, sitting upright just outside the porch door, once again immobile, his eyes fixed on the house door. I found it interesting that he had never once entered a building since we allowed him out of the henhouse, although he climbed them readily enough.

When Manx took his meat into the maples, we went to fetch Snuffles and the dog, discovering them fast asleep on the front seat, cuddled together. They woke up quickly when the door was opened, but they were docile, quite ready to be carried into the house. Inside, we sponged them and rubbed them with towels, leaving until morning the business of bathing them properly; then we fed them, and while Joan supervised their meal, I went to the barn and fixed more wire over the hole that Snuffles had made. Later we settled them in the bear pen, where, I hoped, they would be content to sleep until morning.

It was our turn next. I showered, dressed in clean clothes, and emerged feeling quite human but hungry as a wolf after a winter's fast. While Joan was having a leisurely bath, I tackled supper, cooking up a storm. Later, relaxed, drinking good coffee and sipping brandy from warmed snifters, we discussed the day's events, seeing the funny side of our journey.

By this time it was fully dark outside and because we were tired, we planned to go to bed early. Indeed, no sooner had Joan finished her coffee and brandy than she elected to go upstairs and climb between the sheets. She left. I stayed downstairs, wanting to put on paper the highlights of our crazy journey.

Minutes later Joan returned to tell me that the pup was wailing his head off and would I please go and see what the trouble was. The walls of our house were sixteen inches thick; sitting downstairs, I had not heard the dog, but Joan, opening a window and no doubt listening for such noises, had quickly become aware of the puppy's distress.

Entering the barn, flashlight in hand, I discovered the little malamute pressing himself tightly against the wire of the pen, still wailing and looking disconsolate. Snuffles, however, had retired and was curled up in the hay. Opening the pen, I scooped up the dog when he ran toward me, checked to make sure that Snuffles was still asleep, and left, taking the malamute to the house where Joan was waiting with eyes agleam with motherhood. Once again, I felt sure, our bedroom was to do double duty as a nursery. And I was right, but first the pup was allowed to trot around, familiarizing himself with the ground floor's contents and sampling such tasty things as slippers, chair legs, and rugs.

Tundra was a purebred, eligible for registration at the Kennel Club; his proper name was about two feet long and quite unpronounceable—I can't even remember what it was now. In any event, I had picked the name *Tundra* before I picked the dog, doing so as much because it had arctic connotations and reminded me of my old friend Yukon as because I liked it.

For me, sled dogs are special. And Alaskan malamutes are especially special: They are rugged, they have character, they are intelligent and loyal, and they fear nothing on four legs or two (except, perhaps, a charging bison cow!). The senses of these dogs are almost as keen as those of the wolf, probably because they are the closest thing to a wolf that it is possible to find within the domestic environment. And like the wolf, they are extremely gentle with those whom they love—except, of course, when they are in a playful mood, when strength, agility, and high spirits tend to override discretion.

That night, watching the delightful little dog with the oversized feet as he thoroughly inspected every inch of the house, my mind traveled backward to the days when I roamed the deep wilderness accompanied by Yukon and the imp of restlessness that lives somewhere inside my being once again was aroused. Sitting there, my eyes following the puppy's movements, I conjured images of the far north, of the wildness of the country, of the loneliness that was solace, of Yukon, standing indomitable on some snow-covered mountain while he tested the odors and sounds that formed part of his particular lexicon of life. They had been good years; indeed, they still were, for they had changed me, molded me, had become indelibly etched in my memory until by now they were a part of my innermost self, even to the extent of offering me a refuge into which I could mentally retreat when the demands of my present world became excessive. Joan always knew when I took a mind trip with Yukon; she called such retreats my "wild goose time," for when the big Canadas fly north in the spring, and again when they head south in the autumn, my restiveness becomes pronounced and can only be appeased by a *real* journey into the wilds. Tonight was no exception.

"The wild geese are flying again, are they?" Joan asked, coming to perch on the edge of my armchair. It was more a statement than a question; I didn't need to reply.

Just then Tundra came and plopped himself down at my feet, placing his head between them. Venting a small puppy sigh of contentment, he closed his eyes.

I don't know how long the three of us remained in the living room instead of going to bed as we had proposed earlier, but it must have been longer than an hour. Joan and I just sat in silence, Tundra lay awake, but with his eyes closed, breathing rhythmically; we were three lives attuned and drawing comfort from one another.

The spell was broken when Tundra sat up, ears stiffening attentively, head cocked to one side. Listening with him, we heard it as well, the faint crying of our bear cub, who, deprived of his bedfellow, evidently was suffering considerable anxiety. Joan jumped up, collected the flashlight, and left the house. She returned a few minutes later with Snuffles cuddled in her arms, on his back, his head resting against her shoulder and looking like some placid, hairy, human baby. Now we *all* went to bed; but this time I took the precaution of shutting the bedroom door to prevent our ursine marauder from devastating the kitchen once more.

Morning dawned fine and pleasantly uneventful. Tundra was awake, but stretched out on the floor, and Snuffles was asleep and curled up at the foot of Joan's bed when the first *ding* of the alarm caused me to open my eyes and reach out to cut off the sound. It was 6:30 and full sunup. Amazingly, neither bear nor dog had felt the need to void, which was something that we didn't really believe until we had searched the bedroom quite thoroughly without discovering any hidden surprises.

On this happy note we all made our way downstairs; that is to say, Joan and I used the stairs, our guests were carried, the bear by Joan, Tundra resting in my arms.

Outside, shrieked at by the jays and importuned by the lesser birds, bear and dog each did his thing. While they were thus engaged, I attended to Scruffy the chipmunk, who, veteran that he already was, did not consider the presence of a dog and a bear of sufficient import to delay his peanut demands. I had taken but two or three steps outside the porch when the ragged-looking chipmunk dashed up to me, scurried up my leg, and nosed at my peanut-carrying pocket. With three nuts distending his cheeks, he scuttled back to the ground, ran under Snuffles, and dived into one of his holes.

Meanwhile, Maggot yelled from his plucking tree within the maples and Penny showed herself near the henhouse. Not to be outdone, Beau Brummel, a very beautiful male red squirrel who was one of the "fixtures" in the immediate area of the house, screamed his spite from the safe vantage of an elm, directing his ire at the two intruders. Beau, as we called him for short, had already sired several dynasties of red squirrels and spent his leisure hours in avoiding his wives and offspring, as is the custom among his kind. Indeed, even if he had wanted to live the life of a committed, if polygamous husband, his erstwhile mates quickly would have discouraged him by their subsequent hostility toward the gorgeous redback. They lost no opportunity to scream at him and chase him every chance they got, now and then nipping his hindquarters as he fled ignominiously. For this reason I thought he became somewhat neurotic in between mating seasons.

This morning, because he himself was not being chased by one of his women, he took on the role of property owner, bawling out Tundra and Snuffles for daring to trespass on his home range and reserving some of his spite for me because I had not yet filled the feeders—which I supposed I deserved, for replenishment of the six seed stations had hitherto been my first task in the morning. Bearing this in mind, I called Tundra to me and was delighted when he displayed his superior intelligence and came at once; then I called Snuffles and was annoyed when he took off in the opposite direction, moving in a direct line for Penny, who still waited patiently for *her* repast.

The day had begun so well, too! In any case, I chased after the ambling bear cub, and Tundra, more than willing to participate in any game, chased after him as well, passing me quickly and catching up to his companion before the latter had covered a quarter of the distance that separated him from the skunk. Tundra in this way distracted Snuffles and undoubtedly saved him from a good spraying and saved Joan and me from the awful task of removing musk oil from his woolly carcass. Grabbing the two of them, I retreated toward the house, where Joan was preparing breakfast for all of us; while she was cooking, I got the feed ready for the outsiders and went to distribute largesse.

On this occasion I fed Penny first, not wishing to have the skunk in the yard when Snuffles and Tundra were let out later on, then I placed a pile of mixed seeds near one of Scruffy's dens, filled up all the feed stations, and went to deliver some beef chunks to Maggot, who still whistled his anxiety. Of Manx there was no sign and I congratulated myself for having overfed him yesterday; no doubt he had returned to his own domain by now. Crossing the open space after tending to the hawk, movement in the grass attracted my attention, and a moment later Legs, our resident snowshoe hare, came into full view, her condition proclaiming that she was suckling a new brood of leverets.

Legs was the survivor of a brood of four baby hares brought to us by a neighbor whose dog killed the mother soon after she had delivered her young. The dog also roughed up the leverets, but the owner stopped him and gathered up the very small snowshoes. Three of these died by the next morning; the survivor was christened Legs because of the length of its limbs (a characteristic of all hares and one of the features that distinguish them from rabbits), and because after a cursory, all-too-quick examination, I mistakenly thought she was a buck. By the time we discovered the error, Legs was already pregnant, a condition that she has since regularly attained in due season.

Snowshoe hares, or varying hares, gray hares, white hares, and gray *rabbits*, as they are also called, need relatively little land in order to survive—usually not more than three or four acres; Legs, because she supplemented her herbivorous diet with the seeds that were always spilled from the feeders, seemed to need even less living space. The big hare confined herself to the immediate area around the farm's buildings, using as her nursery a patch of scrubland that lay behind the barn and where each spring she produced between three and five young.

Having been hand raised, the hare was exceptionally tame, but she was cautious, seeming to be able to distinguish between the harmless beings that shared her world, and those that could do her harm. With us, she was always friendly, but if the lynx, or any of the other carnivores were present, she would not show herself. It was probably because of her sense of discrimination that she still survived, for to a lynx the snowshoe hare is a dietary staple.

Although the open space between the house and the buildings, right down to the farm gate, was mostly devoted to lawn, there was plenty of natural cover available to Legs; in addition, she used the buildings when these were not closed, being most at home during rest periods in the workshop, the door of which remained almost permanently open. Watching her now as she hopped toward one of the feeders, I wondered how she would react to Snuffles and Tundra, both of whom, I was sure, would exhibit considerable curiosity when they first noticed her.

Perhaps because I was tired after the previous day's journey, or perhaps because my mind was too occupied with thoughts of the bear and the dog, I became careless soon after breakfast was concluded and again later on, during early afternoon.

Wishing to introduce Tundra to his new home and to give the bear some exercise, I suggested to Joan that we should take them both for a good walk and allow them to meet some of the other animals whose world they were going to share. Accordingly, while Joan tidied away the dishes, I led the pair outside.

I should have known immediately that something was amiss, even before we emerged from the porch. Beau was screaming insults; so were the jays. Scruffy, concealed inside one of his holes, was *cook-cook-cooking* continuously, and even the chickadees were scolding. All these signs spell trouble in the wilderness; they are warnings that proclaim the presence of a predator. But, as I said, I was careless; I didn't pay attention to the language of the forest folks.

It happened quickly. Tundra stayed beside me as we walked toward the north side of the building, aiming toward the large evaporator house that stood at the edge of the maples a quarter of a mile away; Snuffles darted ahead and turned the corner. His rump had hardly gone out of sight when he screamed in agony and fear, a loud, prolonged wail such as I had never heard him utter before, not even when he voiced his distress while crouching beside his dead mother.

As I started to run, I could hear Joan emerging almost as quickly, but before she was out of the porch, and while I was only halfway to the end of the building, Snuffles came dashing back. The little bear's face, neck, and chest were covered in blood, but I was given no time to discover the cause of his injury because he came directly toward me and, reaching my legs, climbed up my clothing, not stopping until he got to my shoulders, there to cling whimpering. Joan had not yet reached us when the author of the cub's wounds emerged to view. Manx strutted from behind the house, stiff-legged, ears peeled back, hackles erect, his eyes slits of fury.

Quickly, I told Joan to take Snuffles into the house and stooped to pick up Tundra even as I spoke. With the pup under one arm I advanced on the lynx, speaking sternly, going directly toward him, the kind of frontal attack that I knew he distrusted. A moment later he had turned himself around and bounded away, crossing the open space between the house and the maples and disappearing within the trees.

Inside, I found Joan mopping the blood from the cub's body. Between us, we cleaned him sufficiently to be able to examine his injuries, finding that Manx's claws had caused three shallow cuts and one fairly deep one. Fortunately, the wounds were not serious, and once the bleeding was stopped and the injuries had been treated, Snuffles forgot about his big fright as he lapped up maple syrup from a saucer.

Snuffles was nothing if not resilient, as are all wild animals. The delights of maple syrup instantly obliterated the hurt he had received from the bad-tempered cat, to the extent that after I carefully had scouted outside to make sure that Manx was no longer in the vicinity, the cub enjoyed himself thoroughly with Tundra during the postponed outing. The two really had become excellent friends, and if the bear, being older and bigger, was inclined to be somewhat rough with the pup, the latter's needle-sharp canines tended to even the odds, so that between romps there often came squeals and yelps of pain; but this was soon forgotten when a new game was initiated by one or the other of them. For two hours the pair ran and wrestled and tumbled, sometimes in the grass, on other occasions in the scrubland, where Snuffles climbed small trees and Tundra yapped furiously at the disappearing rump of his companion.

By lunchtime the two were tuckered out and were put to bed in the barn, Joan and I using the interval to have a quick snack and to discuss the problems that now loomed large before us, all of which stemmed from the presence at the farm of one rambunctious bear cub and one equally rambunctious malamute puppy. There were animals living on our property that would have to be protected from the two latest additions to our ménage, just as there were animals, like Manx, from whom the bear and the dog must be protected. Time would alter the latter situation, for as the two grew, they would either be able to hold their own against the lynx, or the latter would learn to avoid them; in like vein, there were other, less physically formidable animals, such as Penny, that could yet administer swift punishment if they were interfered with. In terms of the inoffensive ones, such as Legs and many others, only extreme vigilance on our part could ensure their protection. Later on, when the bear was old enough to explore the deep wilderness and the dog was sufficiently grown up to respond to training, the lesser beings would enjoy peace and tranquillity. This was not a new problem, of course. We already had housed a number of carnivores whose one aim in life was to eat creatures like the hare, but in each of those cases the time of trial had been temporary and the hunters soon had left to pursue their untrammeled existence in the wild. Some of them came back at intervals, it is true, as Manx did, but they always came to receive food from us, this being easier to "catch" than the tempting yet agile grass eaters. We called these visiting hunters our tax collectors; that's to say, provided we paid taxes in the form of meat, they left our local residents alone. Now, however, the situation was different. The bear would have to remain with us for at least a year; the dog, of course, was to be a permanent resident.

Our major consolation stemmed from the strong sense of self-preservation that is inherent in all animals and which is especially keen among individuals of the prey species. We both had been treated to countless demonstrations of this during the years that we had been offering shelter to orphaned or injured animals, never ceasing to marvel at the speed with which even very young organisms learned to take evasive action and to gauge the extent of the danger posed by some particular meat eater.

When a carnivore is replete, it usually loses interest in hunting, being more concerned with finding a quiet place in which to sleep off its last meal. At those times, the hunter will pass within yards of a feeding prey animal without attempting to charge it, while the grass eater, sensing the mood, contents itself with keeping an eye on the passing killer. This is the general rule, but it is not possible to guarantee that matters will always proceed in this fashion. Animals, like humans, are individualistic; some are especially timid, others can be dangerously bold, yet others can be cunningly bold, taking calculated risks and surviving them because they have learned well the tricks of their particular trade. Imponderables such as these, though worrisome, taught us a good deal about animal nature and furnished the sort of challenge that made an interesting avocation even more interesting.

We did not encourage animals to remain with us beyond the time when they were ready to return to their own world, although some of those that we raised did take up residence on our property and continued to look to us for at least part of their daily food needs; these were in the minority, however, and the majority of them belonged to the prey species. Exceptionally, as was the case with Manx, individual animals continued to visit us at intervals when, I suppose, we *did* encourage them to some degree in that we gave them occasional handouts. Yet we were careful to limit such food offerings to amounts smaller than they would normally require to satisfy their hunger, being able to gauge this because of our experience with the early needs of the animals in question. Now and then, as occurred when I indulged the lynx, we were forced to deviate temporarily from this practice for the sake of expediency.

Much has been said about the inability of wild animals to survive in their proper environment once they have been sheltered by humans, and I am often asked about this when discussing our various wards. There can be no question about the fact that wild animals raised as pets for some considerable time, especially when they are obtained at the helpless stage of their lives, are no longer able to survive if turned loose to fend for themselves. Such organisms have not been trained by their parents; they have never acquired the experience that will enable them to find their natural foods; in addition to this, it is almost certain that they have altered their tastes and will no longer find palatable the foods that their fellows enjoy. As an example: If a wolf pup is taken from the den, raised by humans, and fed cooked meat and domestic dog food it will not be efficient if turned loose and may even refuse to eat its natural prey. Certainly, though it may instinctively chase another animal, its kill rate will be too low to sustain it. Herbivorous animals are less inhibited; their dietary needs are relatively simple and food supplies usually are plentiful; but in this case, they may lose their caution and fall easy prey to the hunters.

For these reasons, our greatest preoccupation revolved around the need to train our animals to live as they were intended to do. This meant that the predators were encouraged to hunt and the grass eaters were encouraged to be cautious. We had some failures in these regards, of course, especially during the early days of inexperience, but the majority of animals that we rescued and later released adapted successfully to life in a wild environment.

Animals are far more adaptable than most people realize. The evidence for this can be seen in any urban or suburban area of the world, where one finds wild animals living side by side with humans and doing quite well as a result. Coyotes, bears, even timber wolves, have got "man's number" and despite efforts made to exterminate them, continue to survive, even increasing in number, while earning their living by filching from man. Squirrels, many birds, groundhogs, deer, and a great number of other species do the same. Indeed, over the millennia of evolution, man has acquired a whole series of camp followers who, though once wild, have now become man-dependent: the rat is one of these, as is the so-called domestic mouse; man has even given rise to his own particular kind of flea, not found on any other creature.

When Joan and I began to care for wild strays we did so untroubled by moral questions; the animals were in need and we helped them; it was as simple as that. Later, we did worry about the *rightness* of what we were doing, but not for long, for we reasoned that, with rare exceptions, those animals that we befriended were the victims of man's interference and would not have required assistance had they been born outside the sphere of human influence. We accepted such victims of civilization, sought to heal them, or raise them, and then freed them in a suitable, wild environment in which they would then be exposed to the same survival challenges that were shared by their fellows; in other words, they would live and die within the precepts of natural freedom. Whether their chances of survival were as good as those of the wild ones of the same species that had never come in contact with Homo sapiens was a question that we found impossible to determine. In some cases, the answer was probably no; in other cases, the animals may well have had their chances increased because they had become wise to the ways of men. Some of our wards did not survive; we had evidence of this. We also had evidence that many of the onetime guests at North Star Farm flourished. Morality, we felt, after giving the matter a good deal of thought, did not enter into the matter. Rather we responded to conscience, and to sympathy, and to the love of life purely and simply; we did our best and were not afraid to face up to our failures and to learn from them.

We were given reason to once again examine this question during the afternoon of the day that Snuffles ran afoul of Manx. In this instance, the bear cub's exuberance and a porcupine's instincts of survival were the root causes of the ensuing to-do.

No wild animal could possibly enter human care at an earlier age than did Spike the porcupine, whom I delivered into the world by cesarean section some three years before Snuffles was rescued.

Apart from the singularity of Spike's birth, his life constantly reminded me of the detrimental impact made upon the natural world by human civilization. For, you see, I killed Spike's mother; I shot her, quite deliberately; and only after I picked up her carcass by one back foot so as to drag her out of the evaporator house did I notice that she was pregnant and close to delivering her offspring. Of course I regretted my hasty act, but by then the .22 bullet had blown the female's brain into atoms.

Later, Joan asked me why I did such a thing, and as I endeavored to answer her question, I found that I was becoming increasingly angry, not just with myself, but with humankind in general. I was the instrument of the animal's death, the executioner, but my own species was at bottom responsible.

In the region where our farm was located (as in many other parts of North America), porcupine populations had increased to the point where the animals were damaging their habitat and thereby doing considerable harm to the regional environment as a whole. The porcupine's diet consists of ground vegetation from spring to freeze-up, when, because of the seasonal death of its food plants, the animal switches to eating the bark as well as the buds of evergreen and deciduous trees. In a wild, balanced state, porcupine populations are kept in check by specialized predators and the damage done to the trees in their habitat is minimal: some trees are girdled and die, but the forest can withstand these losses, and is even better for them in many cases when the thinning out of growth aids the progressive regeneration of the environment. Then too, when a porcupine is feeding on tree buds during winter, it drops considerable quantities of buds that fall on top of the snow and provide needed food for animals such as the snowshoe hare, particularly during years of heavy snowfall. In addition to this, the porcupine, as might be expected of an animal that eats so much vegetable matter, produces copious waste; this returns easily assimilated organic matter to the soil. In this way does nature create balance.

For thousands of years before European civilization discovered the American continent, the porcupine and its predators continued to exist in partnership with the forests of our lands, each an integral part of the overall scheme of life, none being able to threaten its environment because of the many checks and balances that are constantly at work in an undisturbed wilderness. These things changed swiftly, and to the detriment of the landscape, when the fur trade was born. At first the ravages caused by excessive "harvesting" of fur bearers was not immediately apparent. The continent was vast; when one area was devastated, the trappers simply moved elsewhere, and the despoiled land was allowed to recover naturally. But as the centuries passed and more and more people occupied more and more land in North America, the imbalances started to become obvious.

Before the first half of the nineteenth century was spent, John James Audubon and Henry David Thoreau, among others, noticed and recorded many of the problems that had developed and were continuing to develop. Millions of bison were being killed; the species was already in danger of extinction. Equally vast numbers of passenger pigeons were being slaughtered, their journey to extinction slated to become a certainty.*

In recent times, the trapping of wolves, coyotes, foxes, lynx, bobcats, and fisher, all natural hunters of the porcupine, drastically reduced, or even exterminated entirely, these predators in many regions that were readily accessible to trappers. As a result, porcupines began to increase and to do damage to the second-growth forests—the first-growth timbers having already been devastated by the early human settlers. Such porcupine overpopulation is particularly troublesome in those fringe areas where agricultural and timber lands exist side by side. It was especially troublesome in the general region where our farm was located, where there existed some wolves, but no cats (until the arrival of Manx) and no fisher. As a result, the porcupine population had increased to nearly alarming proportions.

A neighbor some four miles from our property told me that he had shot eighty-two porcupines on his farm in one winter; several other neighbors also had destroyed relatively large numbers of the animals. In our own case, I had shot five porcupines when these could not be induced to vacate our evaporator house, a large building in which they took refuge during winter and to which they did considerable damage when they chewed its timbers in an effort to obtain the residual salt that adhered to them in minuscule proportions, but which the salt-hungry rodents could detect. In

*The last known individual died in the Cincinnati, Ohio, zoo on September 1, 1914. At its population peak, the species existed by the millions. In migration, vast flocks darkened the sky; in large nesting colonies, the weight of roosting birds broke tree branches. Today it is gone forever. Since 1600, man has caused the extinction of 36 species of mammals and has placed another 120 species in imminent danger of extinction; additionally, man has caused 94 species of birds to become extinct and has placed 187 on the endangered list—these are world figures.

more recent times, probably because so many porcupines had been killed off by the local farmers, the problem had largely disappeared. There were plenty of porcupines left in our area, but these confined themselves to the timberlands, where damage to the trees was not pronounced.

The execution of Spike's mother took place in the spring. I was preparing the evaporator house for the maple-syrup season and, on entering the syrup-storage section, was confronted by the big rodent. We had had similar encounters in the past, when I would take a broom and push the prickly creature out of the door, then seal up the hole that it had gnawed in order to gain entry. But on that particular occasion the porcupine refused to be ejected; time and again, just as I thought that it was about to leave through the open door, it would twist aside, and waddle away to squeeze itself under the syrup-storage tanks or under the taffy stove. In face of such steadfast refusal to leave, and because its presence, practically underfoot, during the time that Joan and I would have to be working in the building could well lead to our getting smashed by the animal's dangerous tail, I decided to shoot it. Had there been more time, I would have live-trapped it, but the weather had suddenly moderated that year and the sap was about to start running, so, regretfully, I went to the house, loaded the .22 rifle, and returned to the sugar shed. The porcupine was squeezed tightly under the stove, but by going around it I was able to shoot the animal through the head, killing it instantly.

When I realized that its refusal to leave was due to the imminent birth of its young, I felt deep sorrow. On the heels of this I determined to try and save the baby which, I was sure, must still be alive, for its mother had expired less than a minute earlier. But there was no time to lose while I ran to get my dissecting kit with its scalpels. Then and there, arranging the dead female on the cement floor of the sugar house and using my pocketknife, I carefully cut into her body, slicing open first the skin, then the stomach wall, and lastly the peritoneum, that membranous sheath that lines the interior of the abdominal cavity and surrounds the viscera. Now I was able to see the young porcupine; it was already in the birth position, still wrapped in its cowl, that saclike covering that surrounds all mammals before birth and which usually ruptures before a fetus is born, *except* in the porcupine, which retains this natal "shawl" in order to facilitate the passage of its baby quills through its mother's organs.

Holding the abdominal walls open with two twigs hastily broken off from a piece of kindling, I reached in gently and began to lift out the fetus; and suddenly I had a baby in my hands! About to slit the membrane, I realized that I had nothing in which to lay the infant while I severed and tied its navel. I put the baby back inside the warm carcass of its mother, stripped off my wool jacket, and spread it on the cement beside the body, then opened the cowl and removed the little porcupine. Now I cut the navel, which began to bleed. On a line above the stove there were a number of clothespins with which Joan would peg out dish towels used to clean our pans during the syrup-boiling time; I took one of the pins and clamped the bloody navel. The baby porcupine began to breathe and to snort; then it coughed, expelling a plug of mucus, after which its lips began to move in quest of food.

I wrapped the little creature in my coat and ran for the house. Joan saw me coming; as she was to do three years later when I carried Snuffles, she guessed that I carried "something alive." Together we cleaned the newborn, rubbing off the remnants of the cowl, the blood, and the amniotic fluid. During all this the baby squirmed, continued working his mouth, and uttered small grunting sounds. Its quills, tiny and yellow, were barbless, quite soft, and still flexible.

By the time the porcupine was clean and almost dry, I removed the clothespin from its navel stump, which immediately started to bleed again, though more slowly than at first. Now I found a supply of sterile surgical thread and I tied it off, something that wild mothers do not do, but which is hardly necessary in nature because, sharp though they are, their teeth do not slice as finely as a keen knife and the young do not bleed seriously. In any event, the mother continues to lick the cut end, promoting coagulation with her saliva. While I was doing this, Joan was mixing the infant's first feed.

Minutes later, scrutinized by its beady, black eyes—which are open at birth in this species—and relaxed in Joan's lap, the newborn began to feed. Already Joan was committed to the little fellow's survival; I was equally committed, but I didn't really expect him to live. He was eleven inches long from the end of his blunt nose to the tip of his spikey tail and, after his first feed, he scaled thirteen and one-half ounces, a surprisingly large baby for a twenty-six-pound animal to carry, even though porcupines, who breed annually, only have one offspring per litter.

At the end of forty-eight hours of almost constant vigil our newest ward, kept in a homemade incubator, was safely on his way to life. His quills had become hard three hours after birth; now they were as sharp as needles, the longest almost an inch long, his woolly, black fur bristling with the yellow spikes which provided us with a suitable name for the infant.

If Spike's arrival into the world was somewhat unusual, his ability to survive in it was even more so. Having delivered him, I knew to the minute how old he was; I also knew a good deal about the biology of his species; nevertheless, when an organism that exactly forty-nine hours after birth gets out of its "crib" and climbs up a man's trouser leg and proceeds slowly upward until it comes to rest on his shoulder, there to grunt softly and to make small wet sounds, even an experienced biologist is filled with a sense of wonder.

I had read about the precocious behavior of newborn porcupines without having before witnessed such behavior, so in a sense I was more or less expecting Spike to begin to show independence during the second day of his life. Even so, his demonstration was astounding! No other mammal that I am aware of enters so quickly into the arena of life. That very day, when taken outside and deposited among a mixture of clover and alfalfa, Spike began to feed himself. When he had eaten enough, he ambled unsteadily toward me and once again climbed up my body to roost on my shoulder and to nibble tentatively at my ear, his quills prickly, but not unbearable.

From birth and throughout his adult life, Spike's nature can be described in one word: phlegmatic. Rarely did he exhibit excitement (and when he did, he showed it by stiffening his front legs, on which he swiveled to present his rear at the cause of disturbance, meanwhile swishing his bristly tail from side to side); with Joan and me, he was always docile and exceptionally friendly and I could pick him up even as an adult, provided I did so *carefully*. He did inadvertently spike me a time or two, but such mishaps were what might be called contact pinpricks, the quills never penetrating flesh to a depth greater than an eighth of an inch. Nevertheless, small and lightly implanted though these darts were, each experience was memorable enough.

I tended to spoil Spike, probably out of a sense of guilt, and as a result he came often to the house and would stroll inside if the door was left open. Joan, fond of him as she was, did not take too kindly to his invasion of her domain because Spike was never house trained and was in the habit of leaving behind a considerable scattering of brown oblong pellets that sometimes didn't surface until a chair or chesterfield was moved for vacuuming.

After they had rested for a couple of hours, Snuffles and Tundra informed us, each in his own way, that they were ready to leave the barn and to once more engage in the business of growing up. It was a nice afternoon attended by a breeze sufficient to discourage the mosquitoes and blackflies, so Joan elected to walk with us as I led the pair of inquisitive youngsters toward our maple woods. Snuffles, it was to be noted, appeared not a whit the worse for wear after his encounter with Manx; except for the dried, crusted blood that decorated one side of his face, the cub showed no signs of his painful experience and the big fright that this gave him.

As soon as we entered the maple woods, the two became greatly excited as they scurried all over the place, sometimes accompanying each other, on other occasions going their respective ways. Tundra, being younger, stayed closer to us; Snuffles tended to range far afield, but came back when I called him by grunting, which imitation of a she bear's "come hither" command appeared to mean something to the cub. Presently we came to a part of the forest where the trees were thinner and the ground cover more profuse; here grew tall ferns, patches of raspberries, and small bushes. It took the bear but a moment to realize that raspberries, even when still green, were placed on earth by the Creator for the express purpose of gratifying ursine appetites.

Testing the first few berries and finding them to his liking, Snuffles thrust himself into the middle of the patch, disappearing from view. A moment later he cut loose with his second wail of the day, following up on the bloodcurdling sound by emerging a good deal faster than he went in. Balancing the marks of the lynx's claws that decorated the left side of his face, Snuffles now sported a tight group of porcupine quills on his right shoulder. It seemed that the bear cub was disaster prone!

Tundra reacted to his friend's distress by scampering to him and sticking his nose against the hurt, whereupon he managed to collect two loose quills that were trapped in the bear's fur; these impaled the very end of the dog's black nose and caused *him* to scream as he copied the bear's race for comfort.

These events took but moments to develop and it is not possible to describe in sequence the many moves that took place thereafter. Suffice it to say that while Joan tried to comfort Snuffles, I grabbed Tundra and easily plucked the quills out of his nose, whereupon the little dog licked off the twin droplets of blood that resulted and thereafter showed more interest in the yelling bear cub than he did in his minor injuries.

Snuffles fought me when I sought to remove his barbs. In doing so, two of the quills that had penetrated right through his flesh transferred their points into my thigh (I was straddling him) and were pulled all the way out of Snuffles's hide. In the end, I had to carry the cub to the house where, aided by Joan and a pair of surgical forceps, I managed to pull fourteen quills out of the little bear's shoulder, a painful and protracted operation that left Snuffles whimpering and cowed. Then I used the forceps on myself.

We had earlier discussed the need to introduce the cub

and the dog to Spike, intending to impress on them the nature of the prickly animal. I had even hoped that we would meet Spike on our walk, but I did not anticipate such a sudden and dramatic encounter. Be that as it may, the lesson that Spike taught those brash youngsters was never forgotten; henceforth the two gave the porcupine a very wide berth—until Tundra was almost a year old, when he began to tease Spike by dancing around him, but never coming into range of the dangerous, clubbing tail.

At the time of the encounter in the raspberry patch, Spike revealed himself only after both bear and dog were picked up. Then, mumbling at the temerity of the bear, the porcupine waddled out of concealment, stared at us myopically for a moment, then placidly climbed a tree, where he curled himself up for a siesta.

Five

Scattered throughout the wilderness that surrounded North Star Farm and living each in its selected habitat were a number of our former alumni, some of whom, on being injudiciously approached by either Tundra or Snuffles, could turn out to be formidable opponents. Most of these animals kept to the deep forests and were unlikely to be intruded upon by either the bear or the dog; others inhabited range nearer home; a minority, like Manx and Spike, visited the farm as the spirit moved them. And there was Penny, of course; she posed no serious threat, yet her choking and noisome spray was quite capable of incapacitating our juvenile explorers and of causing the humans in their world a certain amount of discomfort during the cleanup operation that would necessarily follow such a spraying.

Two of the animals that were of most immediate concern to me in the days that followed Snuffles's double altercation with the lynx and the porcupine were a great horned owl and a white-tailed deer.

Boo, the owl, visited us regularly, accepted a handout, then stayed for varying periods of time. He was fully adult but of indeterminate age when an anonymous, trigger-happy goon with a BB gun injured his left wing seriously enough to cause the owl to come down in the backyard of a small nearby city. There, seeking only to recuperate, the bird was accosted by the homeowner's beagle, a rash and careless canine whose bravado evaporated when one of the owl's feet struck swiftly, impaling the questing nose with its big, curved talons. The ensuing outcry brought the householder, who arrived at the scene after the dog had pulled free of the claws, lacerating his nose even further when the hooks were dragged through tissue. The beagle steadfastly refused to have anything more to do with the bird, which indicated that the floppy-eared dog had at least a modicum of intelligence hidden within his skull. The man, seeing the mess the owl had made of his pet's smelling apparatus, was equally determined not to touch the owl, while the injured bird, being unable to fly, continued to remain in a fighting crouch, feathers fluffed out so that they almost doubled his size, beak clacking a warning. To break the impasse, the homeowner telephoned the police, and in due course two large and sturdy men in blue emerged from a cruiser and sauntered down the garden, feeling, as they explained to me later, rather silly about answering such a call and determined to stuff the bird into a cardboard box and then, because I was known to them, bring it to my office, located in a neighboring town eight miles away. Matters did not turn out quite as planned.

Halting in front of the owl, Policeman One, becoming suddenly polite, allowed his companion to proceed with the business of capturing the bird. Policeman Two, younger and perhaps more brash because of inexperience, did not hesitate. He bent, started to reach down with both hands, then noted that the bird was raising a spread foot and that this came equipped with two-inch curved talons. Policeman Two straightened up quickly, stepped back, and also became polite. Policeman One asked the homeowner if he could use his phone and, upon being given permission, dialed my office number.

About half an hour later I arrived to find the "posse" grouped near the great horned owl, who still showed by his stance that he was prepared to sell his life dearly. Owls, as the beagle and the two policemen had already realized, attack with their claws, not with their beaks, which are reserved for the final, killing blow when a prey animal has been impaled by the talons; but in cases such as the one in which the bird now found itself, the great, hooked beak opening and shutting with audible clack coupled with the fluffed-out plumes lend the owl an intimidating mien.

After having been caught unprepared on several occasions, I had by now made it a habit to carry an empty cage in the back of the station wagon as well as a variety of equipment that might come in handy when an unexpected animal needed care and transportation. I had designed and made a number of these holding pens, fitting them with two doors, one on the side, and one on the top.

Putting the cage on the ground near the owl and opening its top lid, I used a cloth as a decoy, allowing the bird to grab it with one foot; while it was thus occupied, I bent down and picked it up by the body, in this way preventing its claws from reaching me. A moment later it was inside the pen and the lid was closed. This made everybody happy, except, perhaps, the owl, who was still fluffed out and continued to clack his beak.

When I got home that evening, I checked the bird's injury. The pellet had severely bruised one wing, but the bone was not broken, though it was impossible to determine the number of weeks, or months, that would have to pass before the wing was capable of sustaining the stress of flight; until that time came, the owl would be kept in the henhouse. Eight days later, Boo, as we named him, would feed from our fingers and would perch on my gloved hand, a model patient with only one bad habit, a mania for calling at night, his deep *whoooo* notes being repeated in multiples of three, five, or six, as the spirit moved him, but most commonly in a grouping of five: *whoo, whoo-whoo, whoo, whoo,* deep and resonant calls delivered without inflection and occasionally punctuated by a bansheelike wail. This sudden, piercing shriek is generally used by the species to startle a quarry into movement for the purpose of identifying some indistinct, immobile shape lurking in the forest penumbra. Coming as it does without warning in the quiet of the night, the raucous scream has the power to disturb any creature, including a human watcher who is actually expecting it—to this I can personally testify!

Watching from the vantage of a tree branch with enormous eyes that gather the maximum of light, a great horned owl will shriek; from the reaction that this cry elicits it can then determine whether the object of interest is a potential prey animal, such as a hare, or whether it is a larger, dangerous organism, such as, perhaps, a resting fox, or wolf, whose body outlines have been made indistinct by the darkness and underbrush. If the call fails to stimulate movement, then the bird can be certain that it has wasted its energies on a stump, a rock, or some other inanimate thing.

Like most owls, Boo was quite apt to fly during daylight, an event that usually earned for him the spite of many of our birds; they would harass him in flight, or pester him as he sat in a tree—even the tiny chickadees molested the big predator, sometimes actually flying into him, like Kamikaze pilots. On the whole, Boo did his best to ignore this uncalled-for pestering, probably because he knew he couldn't do anything about it in any event.

I have always been puzzled by this behavior of songbirds when they encounter an owl in daylight. Owls do not molest birds, especially not the small songsters, yet the presence of the night hunters during the day precipitates no end of turmoil, sometimes attracting several dozen birds.

However this may be, Boo had recovered the use of his injured wing long before Snuffles and Tundra arrived at the farm and had shown himself disposed to visit us during daylight on fairly frequent occasions. He offered no threat to Tundra, because he never landed on the ground at these times, but Snuffles, with his climbing skills, might well accost the owl while it was roosting in a neighboring tree; in this case, our bear probably would receive another painful lesson. However, there was nothing that I could do to prevent such a confrontation, so I chose to ignore it and hoped for the best.

The big deer was another matter; she could kill either one of the two youngsters, a fact that may come as a surprise to people who probably think that members of the deer family are inoffensive creatures. This is not so. Deer can, and do, give a good account of themselves when cornered by a predator, or molested by any animal or by a human. They attack by virtually stabbing with their front feet, and their pointed hoofs are unquestionably lethal; even wolves are killed or injured at times, and many a dog given to chasing deer has been stabbed to death. This does not mean to say that deer are normally aggressive; to the contrary, being fleet of foot, they prefer to run under most circumstances, and because they have such good hearing and an effective sense of smell, they more often than not simply move quietly away from trouble long before it reaches them.

Experienced deer are well able to detect the presence of a large predator in their vicinity by means of sound or scent; they are by these means able to discriminate between dangerous and inoffensive animals. The young of even predatory species, if unaccompanied by adults, elicit no escape response from a deer, which usually continues to feed, or rest, when the harmless intruders are nearby. But if a wolf pup, or a bear cub, or a young dog like Tundra exhibits curiosity and approaches a deer beyond what the latter considers a safe distance, the placid browser will snort a warning and probably stamp the ground with its front feet, then will approach the impertinent one stiff-legged and slowly. That usually puts an end to the matter, for the young animal recognizes the threat and retreats. If it does not, it is likely to be killed, or at least so severely injured that it will not recover.

My problem was to cause the bear and the dog to recognize the deer as a dangerous adversary, and to this effect I proposed to lead them both into the white-tail's range and there to exhibit signs of concern by stopping often, growling warnings, now and then turning around to run, then advance again in as stealthy a manner as I could manage, hoping that the intellect and the highly developed sense of preservation instilled in both animals would respond to this warning. Eventually, I anticipated, we would see the deer, at which point I would growl, turn around, and run as quickly as possible; if our wards took heed and followed, all would be well, at least until Tundra was fully adult and feeling confident in his abilities to attack or to defend himself. That was something else again; when this time arrived, we would have to restrict the dog, for it is impossible to remove the predator from an animal that was designed by nature to chase and bring down its food.

The deer in question—eventually named Babe because Joan formed a habit of greeting her by saying "Hi, Babe!"—was not one of our ex-wards; she was fully wild and, as far as I know, born in the area of our property. We met her in our maples soon after we bought the farm and, to Joan's delight, she didn't seek to run away. This was by no means unusual and was linked to the way in which we always traveled through the land, affecting a casual, relaxed gait and talking, when necessary, in modulated tones. In this fashion we proclaimed to the wild ones that we came in peace, and they responded. By this stage in my experience, such movement had become habitual with me for a decade or more and had allowed me to study countless wild animals that I would not have seen if I had moved with the stealth of a hunter, or with the abandon of the uninitiated. Joan, who was not, in any event, given to sudden, nervous movements, quickly adapted to what we called our wilderness pace.*

(In this context, it might be supposed that I am doing a disservice to the animals of the wilderness by disclosing to human hunters this means of approaching game. Not so! To walk in peace with nature, it is necessary to *be at peace*. Aggression is a communicable disease and its symptoms are quickly detected by the wild ones.)

Babe, we were to learn a few weeks after seeing her for the first time, had delivered herself of two fawns that spring, but on the day in question she had hidden them in the deep bush; there, relying on their dappled and spotted coats, on their relative lack of odor, and on their ability to remain utterly still, the twins were much safer than moving about with their mother.

Because we walked a lot through our maple woods and because Babe found good grazing and shelter there, we met the deer almost every day; soon we could go right up to her, within touching distance, or we could sit on a log and she

*By contrast, in grizzly country (or as I did when first approaching Snuffles on the day of his rescue), I move noisily, the intent being to signal my presence when traveling through heavy bush. To this effect, I even packed a sheep bell which I tied to my pack so as to advertise my whereabouts. Naturally, when I was seeking to observe some specific animals, or in areas where visibility precluded my surprising some snoozing silvertip, I traveled less noisily.

would approach us, her great ears moving in rhythm with our voices as we spoke to her in gentle tones. Our friendship was a pure wild thing that came to full bloom when she led her twins to us one evening. Pan, her son, and Pomona, her daughter (Pom for short) walked right up to Joan the first time that they saw her and allowed themselves to be stroked, an action I would have discouraged had I been given time for fear that the fawns would become too tame and fall victims of the hunting season.

By the time that Snuffles and Tundra came to live with us, Pan and Pom, both adults now, were only occasional visitors, coming to the maples during early spring to eat the wild leeks that grew there in profusion, and in the autumn, when they wisely remained within the shelter of our trees during the hunting season, there to stay until the guns became silent, for game animals are quick to appreciate the security offered by land that is posted against hunters.

Wild animals can be extremely capricious at times, even those that appear to have developed a fixed pattern of movement. Babe was no less so. It seemed that when I set out to look for her accompanied by Tundra and Snuffles, she was nowhere to be found, a circumstance that caused me to widen my search area during the days that followed and one that brought us into near confrontation with another well-armed ex-resident, though on this occasion it was Tundra who precipitated the argument.

The previous spring I had released a pair of river otters into an area located about two miles from our property where there was a small lake and several creeks. The pair had settled down readily, making their homes on the shore of the lake on the south side of which I kept a canoe for those days when I took my ease on the water while angling for some of the fat perch that teemed within its depth.

Few animals can rival otters for playfulness. They easily adapt to people if conditions are right and are more than willing to fraternize with their human friends, even after they have been returned to their own environment. By the same token, these large members of the weasel family can be formidable antagonists when aroused. They are exceptionally agile and strong, and they have dentures that are positively lethal.

Slip and Slide, as we named them for their love of slipping and sliding in mud, water, or snow, came to us following an unusual and adventurous series of experiences when they were not quite yearlings.

It transpired that a married couple with whom Joan was only passingly acquainted read in a magazine about a set in Manhattan whose in-thing was the acquisition and exhibition of so-called exotic pets. Hot to trot, this uninformed couple wrote to an address in the United States and in due course received a list describing the animals that were available. This outlet offered for sale ocelots, boa constrictors, coatis, armadillos, monkeys, otters, and various other animals. In due course, after paying more than five hundred dollars, the young otters arrived by air in a crate and accompanied by a whole series of documents proclaiming them to have been inoculated for almost every known ailment and including an attestation that they had not been live-trapped in either Canada or the U.S.A., but that they had, in fact, been raised in captivity. Whether this notarized statement was true or not, I have no means of knowing, but, in view of subsequent developments, I doubt its credibility.

In any event, the proud new owners dashed off to the airport, paid the freight owing, pushed the smelly crate into the trunk of their car, and drove home, feeling that now they really were *with it* and that they would be the envy of suburbia when they walked their otters on a leash. Their eagerness was such that the unsuitable cage was opened in their living room.

The moment that the lid of their prison was raised high enough for their sinuous bodies to slide through, Slip and Slide dived to freedom and thereafter turned the prim dwelling of their "owners" into a shambles. Dirty, fearful, hungry, the otters were not in a benevolent mood as they ripped upholstery, bit their erstwhile "master" on three separate occasions, and then, when Mrs. With-it opened the basement door, scurried down into the darkened nether world. There, continuing their rampage, they chewed and ripped things and refused with savage splendor to allow the foolish suburbanites to enter their domain. After the man returned from the hospital, where his wounds were stitched and where he was given an antitetanus shot, the couple decided that they wanted to get rid of their pets. First they telephoned the pet dealer; he was willing to "take them off their hands," but would not pay for them and would not defray the costs of air freight—it seems that otters sold in this way are considered nonrefundable! Next they tried to unload the otters on some of their friends, none of whom would even take them as a gift. Finally, the lady remembered Joan and telephoned her.

Sure, I said, when asked; by all means we would take them. No, I said with equal positiveness, I would *not* pay five hundred dollars and change for them. Why not try the zoo? My last question settled the matter, for Mr. and Mrs. With-it had, they said vehemently, "had it right up to *here* with the *'@-&!!* otters!"

That same evening, armed with a catching stick (a long stick with a terminal noose that can be opened and closed from the handle end) and with two suitable cages, Joan and I arrived at the unsettled dwelling. About fifteen minutes later we left, the two otters still unhappy, but now, at least, confined individually in roomy cages.

Once again the henhouse was pressed into service. There, once they were fed and had bathed themselves in a large, galvanized tub, the otters started to settle down. Slowly, they came to accept us; when it seemed to me that they looked upon the henhouse as home, I took a chance and allowed them to explore the outside, leaving the hatch open.

The land behind their building sloped fairly steeply. At the top of this incline, against one wall of the barn, the previous owner had placed an old bathtub, one of those cast-iron monsters with legs, which have of late become "collector's pieces." For the benefit of the otters I built a ramp from ground to bath edge.

Happiness is a wet otter! Slide, the female, found the bathtub first and lost no time in diving into it and disporting herself in a delirium of activity that soon drew her companion (they may or may not have been brother and sister). Henceforth I had to leave the filling tap open half a turn to replenish the water that was splashed out or removed on the bodies of the otters when they dashed to the slope—which they soon muddied up beautifully—to slide down it like children riding a toboggan.

When winter came, because I felt that they were still too young to fend for themselves, I allowed them to stay on, continuing to feed them every day, but leaving them to amuse themselves, which they did with total commitment, now making for themselves an ice slide that would not have looked out of place on a bobsled run. The next spring I coaxed them both into a large cage, placed this on the farm wagon, and trundled them away to the lake, where they immediately made themselves at home. But we worried about their ability to catch their own food, and to appease these concerns I visited the lake on a daily basis for more than a week. By the second visit, after they had humped themselves over to greet me as usual, they refused two of the perch that had originally come out of their lake and had been kept frozen for their needs while they were in our care. Later, when they had plunged back into the water, I did a check of the shoreline and found more than enough evidence in the shape of fish heads, bones, and skin, to prove that Slip and Slide were eminently capable of taking care of themselves.

The following autumn, after failing to visit the lake for a number of weeks, I discovered that Slip was gone, but that Slide was in excellent condition. As might be expected, we feared that Slip had been killed, or had become ill and died, and we were sad. But that winter, in January, while I was walking across a second lake that lay a mile and a half to the west of our property, its eastern shore actually forming part of our boundary, I discovered otter tracks all over the place. An hour later, standing on the far shore, at a place where water flowing through the interstices of a beaver dam had weakened the ice, I found a hole the edges of which were shiny smooth; around it the snow was trampled and proliferated with otter pug marks. As I stood there, the noise of snapping brush caused me to turn around and to come face-to-face with Slip, who had evidently found himself a larger body of water and had decided to settle there.

The male otter still lived in the area of this lake the next year, but he evidently liked to travel, for Slide, when visited the week before Snuffles was rescued, had clearly become a mother, judging by the condition of her dugs. Perhaps Slip was *not* responsible, but in view of the fact that the nearest otter tracks that I had come across in the territory were made in mud almost ten miles to the north, I tend to think that Slip and Slide saw each other on certain auspicious occasions.

Trying to find Babe and therefore preoccupied with this task and with the need to keep the dog and the bear under constant control, I was not thinking about Slip when we found ourselves near a beaver canal that issued from the lake. Such channels are common in beaver country and can run from a few yards to several hundred, being engineered by the big rodents so that they may travel in safety to and from distant feeding grounds; this particular channel was about 150 yards long and threaded its way through rather thick clusters of alders.

Snuffles was lagging behind me, Tundra trotted ahead, now and then stumbling over some minimal obstruction due to his oversized feet and his still-poor coordination. Spying the water, he ran to it, thrusting himself between tangled alder branches. His rump was still in sight when I heard a sharp, doglike bark that was followed by a low, coughing grunt; on the heels of these sounds came a *yipe* of alarm from Tundra, who shot out from within the underbrush and dashed back to me.

As the pup sheltered between my legs, Slip's nose stuck out of the brush; this was followed by his head, then his shoulders, a look of rancor in his eyes. Seeing me, and no doubt recognizing my voice when I spoke in order to pacify his blood lust, he inched out farther, intending, I believe, to come to me. Just then Snuffles arrived, his sudden decision to seek my company being motivated by Tundra's yelp of panic. While the cub was busy climbing up my back, the otter, lingering just long enough to cast one disgusted look in our direction, reversed himself and plunged into the beaver channel.

Since matters ended so felicitously, I was pleased by the

Left: George, the flying squirrel, who volunteered to become a ward of the author's.

Below: *Blackie*, during late summer, now a "teenage" squirrel.

Beau Brummel, accepting a peanut from the author.

Spike, the Cesarean-birth porcupine, at two months.

Spike, fully adult and displaying erect quills when disturbed by Snuffles the bear.

Charley, the groundhog, emerges from his home.

Scruffy, on one of his better days.

experience, not only because both Tundra and Snuffles would in future recognize the otter's scent, but also because they now knew that Slip and his kind were not given to benevolence when intruded upon. I was also pleased by the way in which both animals reacted; it seemed that they were at last learning to be cautious and that they recognized the protection that I offered. There was, however, one small fly in my ointment of content: I hoped that Snuffles would soon learn to seek refuge in a tree, his proper shelter, rather than on my shoulders and head; clad as I was in a shirt, his careless claws had left an impression on my person.

On the off chance that we might yet encounter Babe, I steered us toward the maples, but the sought-for meeting was not yet to be. Instead we met One-Ear, the male raccoon that of recent times had set up housekeeping in our vicinity. In those days One-Ear was young, probably no more than two years old, and hadn't yet earned his name. We referred to him merely as "the young male." But after his meeting with Snuffles, he was properly christened.

The business of recognizing individual animals belonging to the same species has often caused surprise in people who are not accustomed to dealing with wild organisms, yet there is nothing astounding about it. As a farmer recognizes his own cattle, and as people get to know their dog among a number of similar ones, so one is able to distinguish between one wild animal and another after meeting them several times. Indeed, I believe it is easier to remember individual animals than it is to remember individual people whom one has met once, casually, for the simple reason that people keep changing their outward appearance with a variety of garments, whereas animals, at most, only change their clothing twice a year; even then, the patterns are always the same, only the color may vary. To a casual observer two beaver of the same size may look exactly alike, but close observation invariably reveals as many differences as those exhibited by humans, not just in the shape of body and the shadings of the fur, but in personality, in movement, and, likely as not, in scars or other accidental blemishes. It is really a matter of observation coupled with the impression that the meeting creates. There are times, perhaps during a cocktail party or some other boring affair, that an individual is introduced to quite a number of people, some of whom will be almost instantly forgotten while others, probably a minority, will be remembered. Some people are made memorable because they share common interests; others may create such an unfavorable impression that this keeps their image alive inside the head, while the forgettable ones have nothing to offer in the way of stimulus. When it comes to animals, at least in my experience, each individual that I have ever met has interested me greatly; only a very few of these have created unfavorable impressions. To date, I have not yet met a truly forgettable animal.

It must be admitted that when one is dealing with fairly large numbers of animals of the same species there are times when the business of sorting them out becomes confusing. For this reason, we made a habit of naming all those wildlings with whom we came into regular contact; all, that is, except for the birds, for to seek to recognize, let alone name, each individual of a flock of chickadees, for instance, is a task too time-consuming to undertake (though I think it can be done). We did, of course, name some of the birds, but only those with whom we had regular dealings on an individual basis. For reasons that escape me, names allow humans to relate with each other as well as with animals and objects. Give a thing a name, and it immediately acquires personality; give it a number, and it is anonymous. That was why we did not assign numerals to our animals, as some biologists insist upon doing in the belief that objective study only can be done successfully in an impersonal way. Bunk!

However this may be, the male raccoon was so busily engaged in digging out wild leek bulbs that he failed to notice our approach until Snuffles, in the lead this time, was almost on top of him. Understandably, the startled raccoon backed up, went into a low, fighting crouch, and snarled a warning, although he did not show himself disposed to attack. Snuffles, after more than a month in our care during which he had eaten like a pig and grown like a weed, was bigger than his masked relative and, now scaling twenty-one pounds, was probably somewhat heavier. Perhaps because the bear was accustomed to playing with Tundra, who was only slightly smaller than the raccoon, Snuffles showed no fear on meeting his snarly "cousin." Instead of backing off as prudence ought to have dictated, he moved closer, whereupon the coon growled all the louder and made a false charge, a sort of warning, the kind of action resorted to often among those wild animals equipped with tooth and claw and which can be interpreted to mean: "Keep away! I don't want to fight, but I will if you don't take off!"

If this interpretation is correct (and experience tells me that it is), Snuffles was either too young or too careless to take its meaning, and when he advanced another step, the raccoon struck at him, seeking to bite his chest. The bite, on later examination, was bloodless, failing to penetrate the bear's thick fur; but this indignity caused Snuffles to lose his rather easily aroused temper. Shaking himself and dislodging his opponent, he sank his teeth into the raccoon's left ear and the latter, who had been obviously reluctant to fight in the first place, snarled loudly, his utterance containing a mixture of pain, fear, and rage. At the same time, he shook himself violently while backing away, which actions caused the bear's sharp teeth to slice through the ear, clipping off cleanly the top third of it. That's how One-Ear got his name.

The entire action could not have occupied more than a few seconds. Once free, the bleeding raccoon scuttled to a maple and ran up it swiftly. Snuffles was about to give chase, but I was near enough to the bear by this time to reach out and grab him, pulling him down when he was not more than three feet from the ground. One-Ear recovered, of course; such injuries are not uncommon, and apart from lending their owners a somewhat odd appearance, the loss of half an ear does not appear to incapacitate them. Whether this injury was to cause the raccoon to become bad-tempered in later life I cannot say, but the fact remains that One-Ear earned a deserved reputation for being aggressive when he became fully adult.

Although I was to remain unaware of it for some weeks, the meeting with One-Ear marked the beginning of the bear's independence, a slow transition from quickly aroused fear that was often contradicted by the kind of scatter-brained behavior that led him into trouble with Manx and Spike to intense curiosity leading to daring, but purposeful, exploration of his environment. He was to remain cautious—a trait without which few wild animals survive for long—but he soon thereafter lost his deeply ingrained sense of fear as he gradually widened his area of exploration during solo expeditions that kept him away from home for several hours at a time. As matters developed, Snuffles and Babe did not meet in my presence, but it is almost mathematically certain that they must have met at some time or another, for each spent a good part of the day within the shelter of the maples. If such a meeting did occur, it produced no untoward results for either party.

Tundra, on the other hand, continued to show insecurity when he was away from us, a not unusual circumstance for that particular breed of dog and one that is shared by wolf pups. The little malamute was daring enough when either Joan or I were with him, but he was quick to scamper back to us if faced by some unknown object or animal.

About three weeks after Snuffles amputated the raccoon's ear, Joan took Tundra for a walk in the maples, and the meeting with Babe that I had sought to contrive took place accidentally and happily. The pup evidently scented the deer before Babe was close enough for Joan to see her, while he was running about twenty yards ahead of my wife. Becoming aware of the other's presence in his vicinity, Tundra scuttled back, staying next to Joan as she walked on. A few minutes later Babe came into view; she too, was aware of her visitors but, apart from raising her head and staring at them, she displayed neither fear nor aggression. The pup dared to move closer to her, but when Joan called him, he returned, at which point she fastened the lead to his collar and led him to within ten feet of Babe. The deer snorted, struck the ground once with each hoof, then retreated a few steps and started to eat; Tundra jumped backward, taking refuge behind Joan.

After Joan told me about the incident, I was forced to realize that the order displayed by the wilderness is more deliberate than haphazard, despite the presence of those animals that hunt to live, and of those that are designed as victims.

There is certainly daily loss of life, but, on the whole, animals live side by side in remarkably pacific ways if one considers the number and diversity of organisms that are found in any given area of a wilderness habitat. If natural predation were the bloodthirsty affair that popular opinion would have it be, the wilderness would self-destruct in quick time; the very fact that life in the absence of man has existed for millions of years and continues to exist today in those places overlooked by humanity's teeming billions attests to the efficiency of natural survival.

In the area of wilderness surrounding our property there lived in a balanced state such diverse creatures as wolves, deer, and a scattering of moose; raccoons, beaver, and foxes; bears, coyotes, and porcupines; otters, mink, and weasels; hares, mice, voles, and squirrels. In the trees and the skies and on the ground lived thousands of birds, some of which fed exclusively on insects; others survived on warm-blooded prey, while still others alternated between feeding on berries, buds, fungi, and insects. Some specialized in forest living; others were aquatic; many waded along the shores of the waterways or in marshes. All these creatures, and more, were cohabitants of the wild land that fortune allowed us to observe, an area encompassing some three hundred square miles of uninhabited country in which, by dint of patience and quiet living, Joan and I had become accepted even by those of its denizens that always had lived there in a free state.

Time and again we were treated to demonstrations of peaceful coexistence between the species. There was the day when my wife and I sat silently on a hill overlooking a fairly large lake that lay some seven miles north of our property and watched a hunter and its usual quarry at peace with each other. We had set out at first light of a lovely spring morning, and had walked along an ancient trail that twisted and turned through a mixture of terrain cloaked by profuse and diverse plant growth. For the first half hour of our trek, we walked through rocky land on which grew isolated pines and spruces and where mosses and ferns and berry bushes grew in companionable splendor; then the land dropped and we had to wade through marshy places and cross ponds by stepping along the edge of beaver dams, our vision restricted to no more than a few yards ahead because of the multitude of sapling poplars that competed for growing space. Two hours from home we penetrated into the gloomy, yet fascinating, understory of a dense forest of black spruce, a veritable cathedral of a place where lances of sun gave the illusion of light shining through stained glass until, as suddenly as it began, the evergreen cloister ended and we stepped onto scrubland, a place where there was just enough soil to sustain mixed species of trees until these grew too tall for the tenuous hold of their roots. The fallen ones littered the ground, but others, younger and light hungry, pushed upward between the sere branches of their predecessors.

For five hours we padded through the wilderness, stopping often to observe and to be observed, or to admire the perfection of a small flower and to marvel at the waxy, artful simplicity of a mushroom. And we collected things: a piece of light gray sandstone that had been dragged out of the soil by the roots of a falling tree and was trapped between two of these tenacious creepers—a piece of ancient coral, turned to stone millions of years earlier, adhered to one side of the fragment of metamorphic rock; the bleached skull of a snowshoe hare, an effigy with teeth and fine etchings that showed how each piece of head bone was intricately joined to its neighbor; the larval skin of a cicada, delicate and empty, testifying to one more marvelous design, for its tenant had changed from a grub that had lived four years under the soil into an insect, the thorax of which housed a soundbox that produced the insistent, vibrating *ziiinging* sound that filled the late summer wilderness.

When we at last reached the confines of the lake, content, but in need of rest and refreshment, we settled on a sun-warmed rock and munched our sandwiches and drank our iced tea and then gave ourselves up to the contemplation of the wilderness. A huge flock of Canada geese suddenly filled the blue above our heads, their eager, discordant yet tuneful voices proclaiming once again the triumph of the sun.

Across the lake, at a place where forest ended abruptly before a flat, rocky "beach," movement attracted our attention. A dog fox came into view, sniffed inquisitively, then padded over the warm rock to dip its silken muzzle into the lake. After drinking its fill, the fox sat on its haunches, scratched its neck lazily with one back paw, then yawned. While it was still sitting there, seeming to be enjoying the day as much as we were, a snowshoe hare hopped into view no more than twenty feet away from the rufous predator. The fox turned its head, watched the hare for a moment or two, then faced front again. Each animal ignored the other, the fox, obviously sated, the hare somehow aware that its archenemy had already eaten. When each of these wildlings had returned to the sheltering forest, Joan stretched out and instantly went to sleep. I got up and threaded my way to the shore.

As I stepped close to the water I startled a frog, one of those little spring peepers barely one inch long; it jumped into the water and paddled toward an area of marshy growth some ten feet away. Beneath it, at one point, the brown body of a medium-sized rock bass drifted lazily upward. The fish eyed the frog, perhaps considering a meal, then flicked its tail and disappeared. The peeper made it to the shelter of the weeds.

Six

The rate at which needy animals arrived to find sanctuary with us was heavily influenced by coincidence. As a result, the number of residents on hand at any given time fluctuated considerably. Sometimes we might have only one or two organisms to care for; on other occasions as many as a dozen (once we had to deal with twenty-two) might turn up within a week or ten days, the pattern being quite unpredictable and often causing logistical problems. Because of the preponderance of young born in spring, this time was usually busiest, but in view of the fact that the vernal season in the wild often contradicts the tidy and rather arbitrary dates ascribed to it by human calendars, it was always difficult to estimate its advent and duration; in addition to this, the variations in mating and gestation times that exist among animals of even the same species, combined with such factors as food availability and the mortality rate among adults (especially among nursing mothers), added considerably to an already imponderable problem.

Once we came to realize such things, we stopped looking for clues that might allow us to predict the start and finish of a "busy season." Instead, we sought to prepare ourselves, our equipment, and food supplies to deal with the unexpected, considering surprise to be the norm. Thus, when Joan and I spoke of spring in this context, we referred to a period that could commence as early as the first week of March and terminate as late as the end of June or even the beginning of July, depending on weather conditions and the vagaries of fate and of the animals involved.

We were not therefore taken aback, or unprepared, when in one day during the end of June three new boarders arrived at North Star Farm. One of these, a very young shrew, was found by me during an early morning, before-breakfast stroll; the other, a young weasel, was waiting for me at the office; while the third, an adult American bittern, was brought to the farm by its rescuer.

The weasel, by virtue of the fact that its youthfulness serves to point out the folly of generalizing when seeking to describe the ways of wild animals, nicely illustrates our inability to estimate the seasonal, possible arrival times of our wards. Fang, as we named him for quite obvious reasons, should have been born during April, according to accepted mating and gestation estimates quoted in most reference works. But it seemed that the weasel's mother and father just hadn't read any of my books, for Fang was no more than six weeks old when he ventured into a garden and was molested by a miniature poodle. Though only five and a half inches long from tip of nose to end of tail, Fang was *muy macho* and, like all the members of his family, equipped with extreme agility, inordinate scrappiness, and a short fuse. The questing poodle was bitten on one paw; this instantly caused the dog to retreat vociferously, *ki-yiing* pain until its master came to investigate.

Fang, caught in the open and instinctively knowing that to turn your back on an enemy is asking for trouble, stood his ground, his sinuous little neck held high, his hindquarters low, tail at half-mast. His unblinking, ferocious, button eyes fixed themselves on the investigating human, held the latter's gaze, and, finally, intimidated the man, who retreated to a garden shed and there took hold of a glass jar. Putting on a pair of garden gloves, just in case, and carrying the jar's screw-on lid, he returned to where Fang remained belligerently at bay and managed to put the glass container over the top of the weasel. After some tricky maneuvering, the lid was fixed on the jar, and air holes were punched.

The man was a reader of my column and had telephoned me a time or two to ask questions about this animal or that bird. On his way to work, he stopped off at my office and left the jar with the receptionist, who put it on my desk without seeking to get to know its beady-eyed occupant. When I arrived, the little weasel was curled up in the bottom of the jar; he looked lean, so I asked one of the girls if she would go out and buy half a pound of stewing beef. While she was gone, I got a cardboard box, scrounged some foam rubber from the furniture store next door, and made a more suitable shelter for the *Mustela rixosa*, or least weasel, to which species Fang belonged.

When the meat arrived, I cut some of it into small strips, placing these in a cup partly filled with water, thus providing food and drink for the pint-sized carnivore. Closing my office door in order to encourage the female members of my staff to remain at their desks—all five had come as a delegation to say that if I let "that thing" get away, they would immediately quit work for as long as it took to recapture the brute—I unscrewed the lid of Fang's prison, coaxed him out of the jar and into the box, and immediately offered him a piece of water-soaked beef. He reared his full inches, rested his two front paws against one of my fingers, and snapped up the offering. Again and again I fed him in this fashion until, at last, he merely sniffed at the meat and showed himself more interested in climbing into my hand.

My official sanctum was one of those glass-walled offices, like huge, square goldfish bowls that so many corporations believe allow the boss to keep a constant eye on his underlings; in my case, the glass was so covered with reminders, calendars, and other assorted pieces of paper that anybody wishing to look out, or in, had to approach within a couple of feet of the walls and peer. I looked up as Fang nestled in both my hands to see a face on the other side of each "peephole" in the glass. Seven people were staring into my office, reminding me of a gang of children with their noses pressed against a candy-store window. It was probably coincidence, but the four-ounce warrior in my hands seemed to play to the gallery at just that moment.

First he sat upright, yawned, shook himself; then he spent a few vigorous moments "washing his face" with his front paws. When his facial toilet was concluded to his satisfaction, he scratched himself in a variety of intimate places, after which he washed his underparts. All finished, he curled up in my hands and settled down for a nap. Carefully, I transferred him to the box, putting the lid on.

That evening, I settled Fang in a spacious, landscaped cage that contained several hollow logs of appropriate size, some pieces of heavy bark, sod in which grass was growing, and some toys, including a couple of crow feathers, a small metal bell, and an empty Robertson's marmalade stoneware jar. It turned out to be Fang's favorite retreat.

From the first moment that I came into personal contact with him, Fang offered me his friendship and trust. He loved nothing better than to run around my person, poking his inquisitive little nose into an ear, or investigating the contents of my pockets; toward Joan he showed the same gentleness, though my wife was just a trifle apprehensive when handling him. The only disagreeable side of Fang as far as we were concerned was his weasel habit of expelling some quite potent musk if he was startled, which caused us to be careful about making noise when he was with us or when we were in his presence in the shed in which his cage was kept. This building soon acquired a definitely noxious aroma since Tundra and Snuffles could smell weasel through the door and Fang could smell dog and bear from inside. Whenever the two visited the environs of Fang's shed, the weasel popped a shot of juice, and this, far from discouraging his inquisitive visitors, caused them to become even more interested in him. In the end, I was forced to transfer Fang's cage to the unused part of our house, putting the weasel in the upstairs section, where he was unlikely to be bothered.

Fang was the first weasel that I had anything to do with, and though there were others later on, he stands out as my favorite. Indeed, Fang appeared twice on television in Canada, in each case when I was interviewed following publication of a new book. The first time, when asked if I had any "wild props," I jokingly said I could bring a weasel. The suggestion was accepted and Fang, the ham in him coming out, put on such a show that he actually stole my thunder. I am sure that the second time I was only interviewed because the producer wanted the weasel, by now quite grown up and most unpredictable in the presence of excitement. The show went off all right—from the viewer's standpoint, that is; but I am positive that there is to this day a producer who blanches at the mention of the word *weasel*.

Fang performed well—too well. He didn't like the light, but his insatiable curiosity got the better of him each time he dived into his den to get away from the big floods, so he was forced to come out again to look around. Unfortunately for those of us in the studio, every time he was exposed to the light Fang saluted with a healthy salvo of musk. By the time we were escorted from the studio, it had acquired a strong personality that was to take weeks to alter, or so I was told much later.

Fang was at best six weeks old when he came to us; this meant that his parents must have mated in late summer and that he was born in mid-May. The weasel tribe has a peculiarity: embryonic development is arrested during autumn and winter and does not resume again until spring. For this reason, full development of the fetus takes at least ten months.

The one to twelve babies born in an underground den are toothless, blind, wrinkled, and pink-naked, but by the time they are three weeks old all their faculties, except sight, have developed to the point that the youngsters are actually housebroken, finding their way by scent to a separate part of their chamber that the mother has set aside as a toilet. During the fifth week of their lives, their eyes open; now they are weaned. Soon after that they leave the den and follow their parents on hunting expeditions.

In North America there are found four species of weasel proper: the least weasel; the ermine *(Mustela erminea)*; the common, or long-tailed weasel *(Mustela frenata)*; and a subspecies of the latter found in the southwestern part of the United States, which is also called the bridled weasel because of distinct black-and-white markings that decorate its face. In winter, in the North, weasels change their chocolate-brown topcoats and their creamy undervests for a white robe that is made distinctive by a black tip on the tail (except for the least weasel, which is all white). This black tip has been the subject of much speculation among biologists and naturalists; some have suggested that the black tail has been ordered by nature to alert the weasel's victims and give them a chance to escape, because, proponents of this theory claim, the only thing that gives away the predominantly white hunter as it travels through the snow is the contrasting ebon patch.

Having watched a great many weasels during wintertime, I find this explanation untenable, first, because the weasel is an expert hunter that rarely goes hungry when prey animals are available, and second, because after many hours of observation I believe that the opposite is true, that the black tip, a small, dark patch traveling jerkily above the snow, against which the weasel's white coat disappears, is more likely to be mistaken for a mouse, or for some other small and harmless creature. Admittedly, I, too, am guessing, but I submit that my guess was arrived at only after I had watched the wilderness for twenty-five years and is thus based upon observation, deduction, and logic. One of the most important reasons for my rejection of the warning idea is the weasel's musk. If anything is capable of warning a prey animal of the presence of the weasel, it is the small hunter's musky odor, which clings to it even when it has not discharged its scent. If *I*, with human olfactory equipment, can smell a weasel at a distance of ten feet, a prey animal, with its sensitive nose, must be able to sniff out the hunter from a much greater distance. Yet the weasel continues to find prey; he is rather good at doing so, in fact.

Weasels are found in many parts of the world and no less than thirty-six species and subspecies have been counted. In addition, the weasel has "cousins"; among these are the wolverine, skunk, otter, mink, badger, marten, and fisher.

Because weasels are fast-moving, nervous animals, they use up a lot of energy during the course of their everyday affairs and as a result have an appetite out of proportion to their size, requiring them to eat one-third of their body weight every twenty-four hours. If they fail to do so, they cannot survive. It is this need that has caused the species to be regarded as bloodthirsty, an undeserved reputation that has given rise to many tall stories about the animal's lust for the kill. In this regard, I am always amazed to discover that whenever man finds an animal that can hunt more efficiently than our own species, this hunter is immediately called *vicious, bloodthirsty, ruthless*, and many other like names. The fact is that weasels must eat more than most animals in order to survive, but because of their size they consume far less than an average household poodle. As for being vicious, Fang and several others of his breed that I have befriended gave the lie to this adjective. I don't mean by this that weasels are the ideal pet; I do not think that *any* wild animal should be kept under those terms. But Fang, from the first moment, proved himself intelligent, docile, and affectionate. He could have bitten Joan or me during the course of countless handlings, but never once did he do so, even when excited enough by other influences to discharge his admittedly noxious scent.

Predatory animals who dare to sink their teeth in any of man's domestic stock are, perhaps understandably, put on the hate list. And, in fact, when a weasel finds itself inside a henhouse, it usually kills in excess of its needs, giving rise to the animal's undeserved bloodthirsty reputation. We talk of clumsy people as being "like a bull in a china shop," and we explain a tempting situation as being akin to "a kid in a candy store." Well, anybody who allows a bull into his or her china shop deserves all the breakages that will follow, but I wonder how many people have ever seen a bull doing its thing with cups and saucers? Yet this cliché points to the domestic bull as an example of clumsiness. In similar vein, it would seem to me that no child should be allowed the freedom of a candy store, because it is certain to gorge on goodies until it makes itself awfully ill. Thus, when a weasel enters the poultry house and finds itself surrounded by big, stupid (compared to wild animals) creatures that cannot get away because of the walls of their coop, it emulates the kid in the candy store. Such bonanzas of food do not occur in the wild; when they occur in a domestic environment, the weasel, or any other predator, for that matter, cannot contain itself. Instead of cursing the weasel, it might be well to ensure that the henhouse is made tight against the little animal.

The tiny shrew that I rescued on the same day that Fang was brought to my office is another good example of so-called bloodthirsty animals. Shrews, who belong to the family Soricidae, have been considered vicious for centuries, but, here again, the reputation of the species is undeserved.

The shrew burns up energy at a rate even faster than the weasel, but because it is so much smaller, it must eat almost constantly in order to stay alive. This means that it consumes *its own weight in meat every three hours* (on the average). This sounds like a lot of eating, of course, but it isn't really; shrews weigh between one-fifteenth and four-fifths of an ounce, depending on species. The smallest measures barely three inches from nose to tail tip and the largest is some six inches long. If, for example, a shrew weighing half an ounce and measuring three and a half inches in length eats an amount equivalent to its own weight every three hours, its total consumption over a twenty-four-hour period will amount to only four ounces.

There is no denying that shrews are feisty little critters that will defend themselves valiantly and are equipped with what amounts to a secret weapon concealed in their salivary glands and consisting of a poison, akin to the venom of a cobra, the purpose of which is to aid the shrew in overcoming animals bigger and stronger than itself. The poison acts on the nervous system, but because it is not injected full-strength into the bloodstream, but is diluted with saliva when the little hunter bites its prey, the venom is not as powerful as that of the snake.

Interestingly, our forebears of two hundred years ago were aware of the shrew's poison, but in comparatively recent times biologists dismissed the story as pure mythology, until one of them was bitten by a shrew and suffered the attendant pain and swelling from the wound. But upon drying and grinding into powder the tiny salivary glands from one animal, then injecting this powder in solution into laboratory mice, it was found that there was enough poison available to kill two hundred white mice!

Some shrews are more poisonous than others, the short-tailed variety being, it appears, the possessors of the most potent spit, a fact that I recalled with vivid clarity on the morning that Tame first drew herself to my attention. The tiny carnivore had fallen into a pail half-full of water and was struggling feebly, already more dead than alive. She was two inches long, but half of this measurement was taken up by her tail; she had white feet, a very pointed nose, and beautiful, soft, blue-gray fur. The little shrew could have been easily mistaken for a baby mole or a young mouse, but having been familiar with the species for some years, I knew what sort of youngster I was about to rescue with a bare hand as I reached into the pail and quickly grasped the half-drowned little beast by the scruff of its neck, getting no more skin between finger and thumb than what one might describe as a generous pinch.

Holding it so that it rested on its back in my right hand—but still captured by the scruff—I noted that it was a female and that she had swallowed quite a bit of water. One does *not* offer mouth-to-mouth resuscitation to *Sorex blarina*, but I ventured to poke her stomach with my left index finger, whereupon she opened her mouth and squirted about half a teaspoon of water over my hand; the second poke produced a little more water and a noticeable intake of breath followed by a small gasp. With a handkerchief, I started drying the small body, but when her chestnut-red fangs slashed into the linen, I decided that she was now sufficiently recovered and that her further treatment could wait until my fingers were covered by something more protective than skin.

Before we reached the house, the shrew started to screech, a spiteful little cry that was accompanied by much kicking and struggling, proving that at least this particular shrew had marvelous powers of recuperation. Twice she almost wriggled free of my grasp, and I was thankful when I finally got her into a suitable cage. Exactly eight minutes after being rescued in a half-drowned condition, the shrew gulped down a piece of raw meat about the size of a hazelnut; half an hour later she ate a second piece, and an hour after that she consumed two more. Now her belly was bulging for a different reason. Now, too, the seemingly savage little creature had become quite tame, accepting the last two pieces of meat from my fingers and, much as the weasel was to do later that day, placing one of her dainty, miniature paws on my index finger. Her tractability immediately suggested a name: tame she was, and Tame she was christened then and there.

When born, shrews are scarcely larger than a honeybee. A female may deliver herself of as many as ten, or as few as three, the average litter consisting of six or seven bee-sized, naked, blind, and wrinkled beings that, about one month later, are turned out of the house so that they may earn their own living. Many succumb, of course, either being killed by any one of many predators or through accidents similar to that experienced by Tame. Because they are such fast-moving, continually active little animals, shrews live a short life. At twelve months they are positively doddering, if they reach that ripe old age. It was this characteristic that prompted me to keep Tame instead of releasing her when she was able to function on her own. I guessed that she was about five weeks old when I found her and I wanted to determine her life-span for myself. For this reason she became the only wild animal that was kept by us in captivity on a permanent basis—permanent in this case turning out to be ten months and five days from her rescue, at which time she died of advanced senility, her body and organs revealing severe atrophy when I autopsied her cadaver.

Nevertheless, Tame had a full life, as shrews go. At first she lived in her cage, where she spent three hours "hunting" and three hours resting on a continuing basis, each period being remarkably regular; one could almost set one's watch by Tame's work-and-rest schedule! We fed her lots of red meat, but I also kept her supplied with a generous quota of bugs, grasshoppers, slugs, ants, and any other insect I could catch or that committed suicide by entering her cage unasked, many of which did at regular intervals, for I kept her outside in an area that was fenced off with large-mesh chicken wire and roofed with plywood to keep Snuffles and Tundra from intruding on her. The wire was sufficient to discourage Tundra, but it would have been torn up quickly by the bear if I had not resorted to a nasty trick.

I wired an electric cattle-fence generator to the chicken wire. Snuffles discovered the new pen and had to check it out; he stuck his nose against the innocent-looking mesh and immediately sought to turn himself inside out in his eagerness to remove himself from its vicinity. Hair sticking out stiffly, Snuffles wailed and ran, heading, of course, straight for Joan, who had emerged to see the experiment.

The electrical charge produced by such generators is not continuous, but is spaced rhythmically; in any event, it is not strong enough to inflict serious shock, even if the second-long jolt is anything but pleasant. To prove to my wife that the thing was not as nasty as she claimed, I grabbed the mesh myself and endured half a dozen smacks; even so, Joan thought I was mean. The fact remained that Snuffles was to avoid chicken wire from then on, with or without the electric fencer there.

When Tame had grown to her full, imposing length of four inches, of which her tail measured one and three-quarters, I made for her a sort of outdoor terrarium by digging a square hole in the ground, lining its sides with tin sheeting to prevent her from climbing up, and supplying her with a number of natural playthings, such as rotting logs, bark slabs, and grass. To allow her to make her own den, I put a mound of earth in one corner and offered her a good supply of dried leaves and straw. Soon she had tunneled into the dirt and had transferred most of the nesting material into her bedchamber. This terrarium had a distinct advantage over the cage; to begin with it was more hygienic, and it saved me hunting insects, for these committed suicide in droves by exploring the shrew's quarters. She especially loved big, fat, juicy June beetles; once I saw her eat four of these in one sitting. Now and then some foolish frog dropped in to say hello and stayed to furnish a veritable feast. Tame, it seemed, was willing to try anything at least once provided it was made of meat. Earthworms, who often surfaced in the wrong corral, she slurped up like spaghetti *al dente*, centipedes were an hors d'oeuvre, and the juicy larva of flies were her equivalent of Beluga caviar. Once she cornered a large vole, a creature more than twice as big as she was and which supplied her with food for three days and then, by attracting a great many insects, including carrion beetles, further stocked her larder. To this day I think of Tame with amazement and fondness, for with me (Joan refused to go near her when she discovered the shrew's tastes) she was always docile.

Tame was settled in her cage before I left for the office on the day that she was rescued. When I returned that evening with Fang, it was to find that Joan had accepted responsibility for an American bittern that had flown into the radio antenna of a slow-moving car on a local road. The driver was a farmer in our area and although we did not know him, he had heard about us; his wife was a keen bird watcher and she insisted that they bring the injured bird to us.

Joan, after taking a look at the bird's formidable bill, had got the farmer to put her in the henhouse, where she remained, pretending to be a dead stick, until I camc home. But if my wife had not checked the bird to determine the extent of her injuries, she already had found a name for it: Goggles. This was suggested by the way that the bittern's eyes moved, the species being capable of rolling their staring orbs up, down, forward, backward, and sideways.

Goggles, a female it turned out, had not been seriously hurt. When I examined her, in itself a tricky operation, I discovered that her elbow had become dislocated and she had a nasty bruise on her side.

Bitterns and herons are potentially dangerous birds because of their lancelike beaks, which they do not hesitate to use in their own defense. One naturalist I know lost an eye when a heron he was seeking to band pecked at him; the blow was so strong that it came within an ace of piercing the man's brain. For this reason, extreme care should be taken by those who would handle one of these big birds. In my own case, the first thing I did after Fang had been settled was to grasp the spear beak in my left hand while holding the bird's legs with my right. Joan now tied the legs with strips of soft cloth, then, using surgical adhesive, taped the beak shut, leaving a tab of adhesive sticking out on the underside of the bill; to this she tied a length of cord, which I then fastened to the bittern's legs, securing the beak in such a way that Goggles could not lift it high enough to lunge.

Popping the displaced bones back into place took several minutes during which Goggles experienced considerable pain. When the job was done, I immobilized the wing by binding it to her body with a bandage, leaving her good wing free. Now we untied her legs and lastly peeled the adhesive from her beak. The patient recovered, but she was unable to fly until early fall.

In the meantime, about a week after she came to us, she had adapted to her new life and was tame enough to be allowed her freedom, though she was shut in the henhouse at night for fear that she would be picked off by a wandering fox or lynx. It was during her second day of free living that Snuffles and Tundra discovered her presence while Joan and I were still inside the house.

I happened to look out of the window at the moment when Goggles emerged into view from between the barn and the machine shed, strutting like some jackbooted storm trooper on ceremonial parade. As I watched her, Snuffles appeared, bounding like a large, hairy ball and going directly toward the bird; as might be expected, Tundra was not far behind his ursine pal. Goggles, seeing them coming, went into her dead-stick routine, squatting down on the grass and raising her neck and her beak until they were fully perpendicular, aiming at the sky. In this position, her extraordinary eyes were able to watch the bear and the dog, so that when Snuffles got near her and injudiciously offered his rump while he sought to get behind her, Goggles delivered a smart peck on the hairy backside, an action that elicited an immediate scream from the bear, caused him to jump, all four feet leaving the ground in unison, and then, after he had landed again, propelled him at speed to the barn wall, up which he scurried as lightly and swiftly as a squirrel, yelling all the while.

Tundra immediately applied his brakes, but Goggles was by now aroused, and as the little dog turned, he, too, received a sharp jab on the bottom, though he was already moving when this was delivered so that it did not connect as solidly as did the one that had spurred Snuffles onward and upward. Nevertheless, Tundra yelled as he ran for the porch, where he was greeted by Joan, who picked him up and told Goggles that she was a number of uncomplimentary things.

That left Snuffles perched and whimpering on the barn roof, from where he refused to descend because the source of his anguish was once again squatting and seeking to transform herself into a sere reed. I emerged from the house at this point, having been immobilized by laughter while the action outside unfolded. My one regret is that I was unable to capture the scene on film.

When I picked Goggles off the ground and carried her to a more suitable part of her domain, Snuffles allowed himself to be coaxed down from the barn, whereupon he was invited into the kitchen and spoiled once again with a saucerful of maple syrup. At this rate, the young bear was going to *seek* trouble so that he could be rewarded after each clash! But the affair caused neither Tundra nor Snuffles any lasting hurt; Tundra escaped with a dimple where one should not have been; Snuffles's punctured seat soon healed.

Toward the end of July I noticed that the friendship that had instantly developed between Tundra and Snuffles was starting to wear thin. The dog had become clearly jealous of the bear and exceptionally possessive of the house and its occupants. It was impossible to mistake the change that had taken place in the dog's attitude by this time, but I could not determine when the change had begun.

Snuffles was about six months old, Tundra was four months; each had grown considerably, the bear now weighing thirty-five pounds and the dog twenty-two, but whereas Snuffles continued to be cautious, our malamute was becoming increasingly independent and daring, each animal exhibiting characteristics common to their particular species. The canines achieve independence from their parents a great deal earlier than do the ursines, who stay with their mother until they are about eighteen months old.

The bigger they grew, the more vigilant Joan and I had to be, because each animal was now capable of killing the smaller wild ones that abounded on our property. In this regard, the dog was the greatest offender, coming from a long line of chasers, whereas the bear, being omnivorous, spent much of his time away from home browsing grass and searching for fruits and berries and insects, especially ants.

During the first week of July, Tundra's love of the chase earned him a good dose of Penny's musk, but despite the fact that he unselfishly shared the stink with us when he came bolting into the house after the fact, the event was beneficial because from that day Tundra avoided skunks with the same dedication, taught to him by Spike, with which he kept clear of porcupines. He continued to find Legs, the snowshoe hare, totally irresistible and, regardless of scoldings or scruff shakings, would not miss a chance to run after her. Fortunately she was more than capable of showing him a clean pair of heels as she darted away at high speed, always traveling in a circle, as is the habit of the species.

Because of its circular mode of escape, many people believe that the snowshoe hare is not very bright, arguing that by staying within such a relatively small area it is increasing its chances of death. This is not so; indeed, the opposite is the case. By virtually returning to the point from which it was startled, the hare confuses its enemy by laying a round trail that, in effect, has no beginning and no end, the odor of hare being equally fresh at any given point of the circle. Repeatedly I watched Tundra give up after sniffing around like mad while Legs sat still as a statue in some patch of bush and watched the dog run itself into confusion. I have seen foxes and wolves become equally baffled by this trick that nature has taught *Lepus americanus*.

Snuffles, on the other hand, seldom chased anything unless he almost fell over it and then only after he had paused long enough to make sure that it would not chase *him*, so that by the time he tried to take after a creature like Legs, he was hopelessly outclassed. But he did seem to develop a fascination for Goggles, who had to administer two more jabs on his corpulence before he at last consented to leave her alone.

By now Snuffles could be trusted to remain outdoors all the time and to bed down in the barn when he felt in need of rest. Habitually he spent a good part of the day snoozing in his hay bedroom or dozing in a convenient tree, hanging himself over a couple of branches like so much wet washing carelessly tossed over a line. In this he demonstrated a natural aptitude, shared also by the raccoons, and it was not unusual to find the bear having a nap at one level of a tree and a raccoon or two doing the same thing higher up.

Indeed, it was commonplace for a large female raccoon, whom we named Herself for no particular reason, to share with her brood of four the bear's favorite tree, a large sugar maple growing near the farm gate.

Herself was a totally wild raccoon, that is to say, she was never one of our residents, but arrived one morning in May while I was sitting in the sunshine having coffee and toast. Feeling that eyes were upon me, I looked up into a wild plum tree that was some twenty-five yards from where I munched, and there, hanging partway down the trunk, was an exceptionally large raccoon who made a perfect picture because she was framed by a profusion of pink blossoms. Moving quietly, I went inside the house to get a camera, emerged again with one that was fitted with a 200-millimeter telephoto lens, and walked slowly toward the raccoon. She stayed as she was, looking at me. I took several photographs of her before strolling to the outdoor table to break off pieces of toast from the remaining slice on my plate.

Herself watched the first morsel sail through the air; she noted where it had fallen but maintained her fixed position, watching me. When I returned to my seat and pretended disinterest by sipping coffee and turning my gaze elsewhere, the raccoon started to climb down, the sound of her claws scraping against the tree bark giving away her movements. After a few minutes, I looked her way again to see her sitting in the grass munching on the toast, her full and hanging dugs telling me that she was a nursing mother. Finished with the first piece, Herself raised up on her hind legs and stared at me; I tossed her another chunk of bread, deliberately throwing short, so that she had to travel a few yards nearer to me if she was going to retrieve it. She did, and from then on was a regular visitor who eventually fed from our fingers and was trusting enough to walk into our kitchen through the window. She would sit on part of the countertop while she fed on the good things that we put out for her. Her visits, with or without her young, did not disturb Snuffles, but caused Tundra to behave like a spoiled brat, growling, dashing around the kitchen, trying to jump up on the countertop, and actually drooling.

During such performances, Herself was content to watch the dog, being wise in the ways of the hunters and knowing herself to be quite safe. Nevertheless, after several such demonstrations of canine temper, we started banishing Tundra to the basement whenever we allowed Herself to come into the kitchen. It may be that the feud that developed between the two animals in later years was due to the dog's exclusion from the room, which he considered to be a punishment; Joan claimed that it heightened his dislike of the raccoon. I'm not so sure; I think that in an animal like a malamute, even a young one, the love to hunt transcends all things.

Seven

It was late August. Tundra and I had gone for a long walk in the wilderness and were returning home after some six hours of exercise, stimulation, and adventure, the last being almost exclusively experienced by the five-month-old malamute, who started things off in high gear when he put up a dog-fox less than half an hour after we left home.

The dog had grown considerably during the past month or so; he now weighed forty-two pounds, but his size was insignificant compared to his conceit. With colossal arrogance, the young Alaskan clearly believed that he could now master anything on four legs that he might encounter in the wildwoods. Not long after chasing the fox into the deep forest, Tundra took a voluntary swim off a beaver dam during an attempt on the life of a young mallard duck; he scarcely had shaken the moisture from his body when he scented a vole sheltering in a clump of grass, upon which he pounced, the little gray rodent escaping by a whisker. And so it went, from one form of life to another; or put another way, from near miss to near miss.

Seeing that he was still too inexperienced to actually make a kill, I allowed him to exercise his predatory instincts and to work off some of his surplus energy, being rather pleased with the way he was developing. I didn't really want to encourage him to hunt, but at the same time I didn't want to blunt his inherent urges to the point where he would become frustrated and turn vicious; I had seen too many sled dogs go bad because of such curtailment. Just now he was hunting bloodlessly, and that was good; later on, when he had the coordination, power, and experience to follow through with a kill, I was going to have to restrict him, but by then I hoped he would have become a working dog, and this would allow him to get rid of the "steam" that is so characteristic of the breed.

Thinking these things as we turned for home following a different route, I was unprepared for Tundra's sudden spurt only moments after we emerged from the forest and entered a small clearing, but it took just one glance to identify the cause of the dog's charge. Grazing in the center of the open place, a shaggy black bear had his back to us; he was unaware of our presence because the soft breeze was blowing in our direction. But he soon heard the thudding of Tundra's oversize feet.

Caught in the open, the bear turned to face the dog, shaking his big head from side to side angrily and *woofing* loudly to make it clear that he was not in a good mood; at the same time he bounced up and down, keeping his back legs on the ground while raising his forequarters, and in between bounces lowering and raising his head jerkily. That he did not turn on Tundra probably was due to the fact that I interfered, for when an adult bear, and especially a cranky old male as this one appeared to be, resorts to such antics, it means that sane animals (and people, too!) had better recognize the symptoms and discover that they have urgent business elsewhere. But not Tundra!

Behaving as though a four-hundred-pound bruin was something that he was accustomed to slaying every morning before breakfast, the foolhardy young malamute continued his charge, his neophyte hackles raised stiffly on neck, shoulders, and rump, his tail, not yet able to achieve a full, adult curve, raised as high as it would go, the last third of it held in a loose spiral.

It was then that I yelled, clapped my hands, and started running toward the two, making as much noise as possible. Such reinforcements were too much for the bear's cautious nature. He wheeled like a well-trained cavalry charger and ran with ursine speed toward the trees, reaching them in no time and quickly scaling a tall elm. Tundra, not yet able to match an acceleration rate that puts the needle at twenty-five miles per hour, nevertheless fetched up with a thump against the elm while his quarry was still climbing at the midway point. The pup was trying to go up after the bear when I reached him and grabbed him by the scruff of the neck, lifting him off the ground and speaking crossly.

Tundra was not abashed. He licked my face, telling me in his canine way that he thought he was one heck of a dog and that if I hadn't interfered, he would have murdered the bum! I carried him well clear of the bear's tree before putting him down again and we continued on our way, but side by side, because my errant dog was now connected to my right hand by his lead, something he really didn't care for. Ignoring his useless attempts to eat the offending halter—which was made of fine, stainless-steel links—I aimed us straight for home, my mind occupied with the recent incident.

Tundra and Snuffles had not been getting along at all well lately. The dog's jealousy, which first became apparent during early July, intensified as the malamute grew bigger, stronger, and more confident. Inasmuch as canine young develop at a faster rate than bear young, Tundra was now almost as large and heavy as Snuffles, despite the difference of two months in their ages (the dog was born March 27). In addition to these things, the canids mature earlier and are naturally more aggressive.

Our malamute's jealousy of the bear was in part triggered by his inherent pack instincts, the same instinct that causes wolves to seek higher status as soon as they feel confident enough to challenge their immediate hierarchical superiors, and in part by the dog's understandable desire to seek for himself the largest share of our affection and notice. If we had been farsighted enough to anticipate his resentment of the bear, we might have been able to prevent it, but, lulled by the instant friendship that had sprung up between them, we failed to note the signs early enough, so that by the time Tundra's hostility became pointedly obvious, the makings of a lasting feud had developed. Since becoming aware of the problem we had tried to change the dog's attitude by showing him out-and-out preference and by spending a good deal of time with him alone—as I had done this day—but our efforts thus far had been to no avail. Indeed, I reflected as we made our way back to the farm, Tundra's deliberate skirmish with the big bear this afternoon suggested that he had now broadened his animosity so that it included the entire clan. In this he reminded me of Yukon, who, after an initial fight with a black bear during which his mouth was slashed, declared total war on every member of the species that he met. But Yukon was big, weighing 120 pounds when he took it upon himself to rid the world of *Euarctos americanus* and *Ursus horribilis;* he was also half wolf and shared the lupine agility and endurance. Tundra, however, was only a partially grown pup, a precocious juvenile without advantage of weight or experience. These were disturbing thoughts in view of the fact that I could see no solution ahead. Nevertheless, I concluded as we came in sight of the house, Joan and I were going to have to try harder to change the dog's attitude.

The sun was dipping low in the west when we entered the house to find Joan sitting on the chesterfield holding Snuffles in her arms. The bear was partly on her lap and partly on the sofa, his head and shoulders cradled against Joan's side, my wife's thumb in his mouth; he was sucking gently on this comforter, so content that apart from sliding one eye toward us as we entered the living room, he did not budge. Joan smiled, but she also remained unmoving.

My wife's maternal behavior and the cub's infantile acceptance of the comfort that it offered set aside the obvious differences that exist between humans and other mammals. This caused me to realize for the first time that all mammals share certain basic, but essential, characteristics.

Eventually, I was led to conclude that the young of all animals are motivated by the same needs and exhibit nearly identical emotions and reactions during the helpless stages of their lives: infantile needs create emotional responses that are wholesomely honest and immediately apparent; these, in turn, are able to stimulate latent maternal urges. Thus, it seems to me that human females are strongly motivated to respond to appeals made by mammalian infants of almost any species, but there is enough evidence available to prove that far from being an attribute unique to our own kind, it is rather a universal characteristic of the female animal, a latent tendency that surfaces under the right conditions and regardless of species. Thus will a female cat nurse a puppy, or even a very young mouse; in the same way will a bitch suckle a lion cub and continue to consider it her offspring long after the feline has reached maturity. We once had an aging golden retriever, inherited from friends who had to find her a home when they went to live in England; Kim sought to mother every small animal that she came into contact with. It made no difference to her whether the animal was a chipmunk, a squirrel, raccoon, or anything else, or even whether it was young or adult: if it was small and furry, she would try to care for it.

This was long before the arrival of Snuffles and Tundra, of course, and Kim had succumbed to age and cancer less than a year after coming to live with us, but her love of all things made her unforgettable. She was overweight and not given to much dashing around, but she would lie on the ground in front of our doorway and watch the birds and squirrels and chipmunks; they, in turn, quickly came to accept her, to the point where some of them used her as a sort of springboard, not bothering to go around her, but hopping atop her and launching themselves from there. And Kim, it seemed, was glad to be of service; she would look at them almost adoringly, the tip of her tail wagging gently each time she was bounced on.

Because Tundra was about to strangle himself in his eagerness to go to Joan and seek her affection while ensuring his rival's ouster from her lap, I picked him up, took off the lead, and carried him to the chesterfield, sitting beside my wife and the bear. Thus the magic moment was dispelled in a peaceful way especially when Snuffles stretched, yawned prodigiously, and slipped to the floor, there to scratch his belly a couple of times.

Joan got up and led the bear to the door, opened it, and allowed him to go outside. As she did so, I had to hang on to the dog, who either wanted to get at the cub or else claim his share of Joan's attention. The bear no longer being available, Tundra ran to my wife.

After supper, with the dog's head resting on my left foot, I thought about the affinities observed in young animals of dissimilar nature, and particularly on the one trait that I had found to be common in all of them. This can best be described as the way in which new life is naturally programed to be selfish—to seek comfort and sustenance first and foremost and to form a quick attachment for the being who offers these necessities, no matter what the relationship between the two creatures may be.

In this context, I was reminded of Squirt, a young raccoon that I had rescued two years earlier after he had been injured when a bear ripped open the entrance to the female raccoon's denning tree and killed all but one of its occupants.

The forest drama took place in late spring and came to my attention after I had set out on my customary before-breakfast stroll clad only in shorts and running shoes. I had been walking for perhaps ten minutes and found myself within the shelter of a dense patch of mixed forest where grew isolated cedars interspersed with poplar saplings, maples, and a few elms. One of the cedars was old, large, and gnarled, its trunk hollowed by disease and by the efforts of succeeding generations of raccoons that had found shelter and warmth inside it in winter and had used it as a nursery in spring. The entrance to the well-used den was located about twenty feet up the trunk, at a point where three heavy limbs triangulated before rising toward the light. Lower down, no more than three feet from the ground, the trunk had split, opening a two-inch-wide gap that traveled vertically up the tree for about four feet. Joan and I had often watched raccoons scaling up the inside of the trunk, pausing to peer at us through the slot. On that morning I was curious to see if the tree had again been used as a den, so I walked toward it.

The irregular, splintered hole had been made by some big and powerful animal at a point two feet from the ground. Only an adult bear could have carved its way into the tree in that fashion, getting a start at the two-inch split and tearing out great slivers of yellowish sapwood. That the marauder had but recently quitted the area was evident by the glistening sap that oozed from the tree's wounds, but if further evidence had been required this was presented by several large, flat-footed prints made in the mossy, damp understory, into which water was still oozing as I knelt to examine them.

Other evidence was available to proclaim that this was the work of a black bear: pathetic bits of raccoon fur, one small, black foot equipped with humanlike digits, about a third of a young raccoon's ringed tail, patches of blood-soaked skin. Inside the broached den, sticky-wet pools of blood told their own story.

That part of me that is human and sentimental recoiled from the slaughter and became angry; the naturalist side of my nature understood that a hungry predator had followed its nose to the tree, forced an entry, and assuaged its appetite with the food that it found inside the den chamber.

No matter how often I find the leftovers of predation, my emotional side seeks to contradict the clinical part of my mind, but training and years of field study oppose sentiment, arguing to such effect that it becomes impossible to deny the rightness of the scheme of life, which is premised on death, my own included; awareness of our mortality is perhaps the main reason why most humans take the part of the "underdog."

On that particular morning, however, my inner struggle was interrupted by the sounds of slow, stealthy movement coming from nearby, noise that was so distinctive that even without looking for its author I knew that at least one of the raccoons from the den had survived the bear's attack.

When seeking to avoid detection, raccoons virtually

A young vole, or lemming mouse, found by author after rains flooded its burrow.

Crazy's offspring, held by author, after removal from toolbox where mother made her nest.

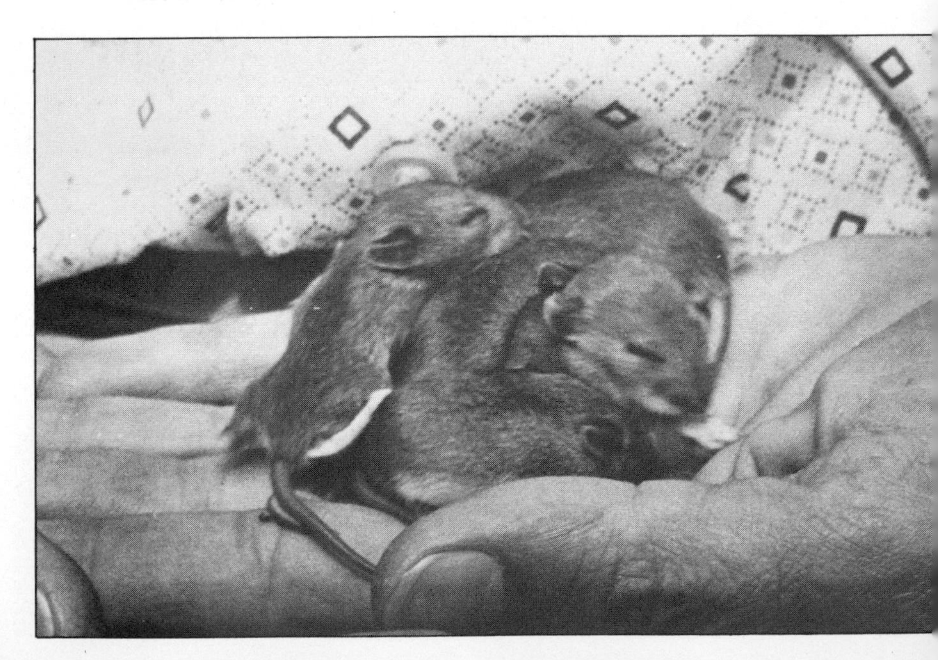

Weavil, being introduced to Tundra.

Weavil, as a young adult.

Legs, in winter, posing after a wash.

Young snowshoe hares held by Joan after their mother, Legs, delivered them in vegetable garden.

Penny, the skunk, at work on Pablum and Olac lunch.

Left: *Tame, the shrew, devouring a frog as big as herself while a salamander hovers nearby unaware of its danger.*

Below: *Fang, in the author's hands.*

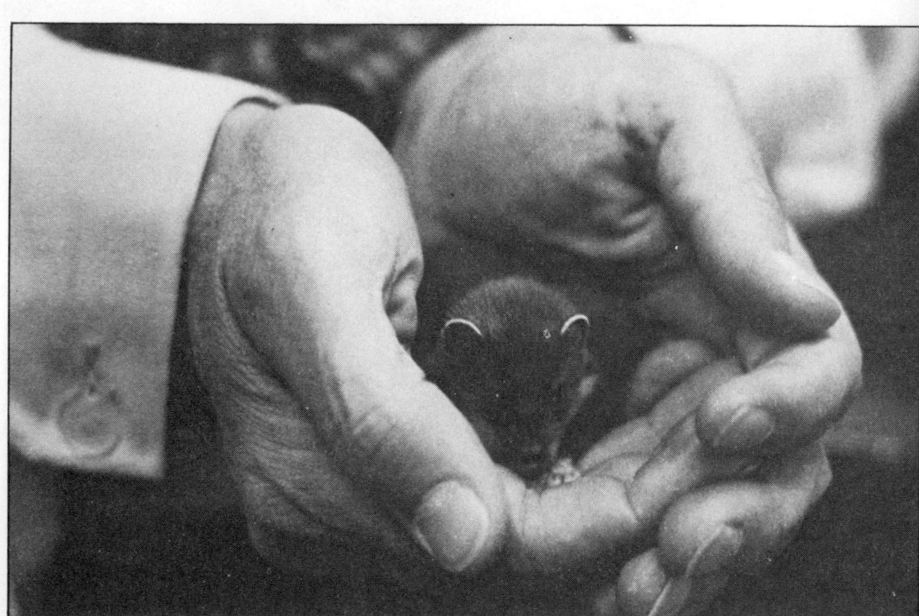

Above: Crazy, the white-footed mouse female who dared to steal food from Manx the lynx and got away with it.

Left: *Manx, the lynx, almost recovered from his injury and now more friendly to humans but yet looking upon them with obvious disdain.*

climb on tiptoe, going very slowly and carefully and making barely a whisper of sound. With the practice of experience, it took me but a moment or two to spot the animal as it sought to find concealment high up in the sparse branches of a sapling poplar that was no more than ten feet from where I was standing. It was a young raccoon, its size and the season suggesting that it was only seven or eight weeks old; it had placed the slim tree trunk between me and itself, but its body overlapped both sides of the poplar, revealing one back leg that was bleeding. Walking around the tree to get a better view of the coon, I noticed a number of bloody daubs on the light green bark.

The poplar was no more than four inches in diameter at the butt; the raccoon was about twenty feet up, near the top. My next moves were reflexive, triggered by my concern for the little wildling and no doubt also motivated by the bout of sentimentality that I had experienced when first confronted by the ravaged den. I wound both legs around the trunk and began to climb. So did the raccoon, moving himself higher, but still at that slow, measured rate.

Minutes later I was near enough to reach for the snarling orphan, a male who, though injured and obviously terrified by the bear's onslaught, was pretending that he was going to fight for his life. Securing a tighter grip on the skinny tree trunk with my legs and left hand, I stretched my right arm and took hold of the raccoon by the back of the neck, lifting his clutching paws away from the trunk and bringing him toward me, then holding him against my body. It was then that I realized that I was naked from the waist up!

The orphan, as I had expected, became quiet; he pressed himself against me, his claws aiding him to get closer. Now I became aware of the full extent of my predicament. Whether going up or down a tree of such meager girth it is necessary to use both legs and *both hands*, but there I was, too far from home for Joan to hear my calls, holding a very unsettled young raccoon against my naked chest, and clinging to a skinny poplar with only three of my limbs.

For some moments I stayed that way, holding the raccoon against me as gently as possible and trying to find a solution to the problem. The one that immediately presented itself was not to my liking, but by the time that legs and left arm began to ache in protest of the strain that they were undergoing, I was forced to put the plan (if such it can be called) into effect. Hitching the raccoon higher up my body, so it could cling with its front paws to my shoulder, I let go of him and went down that tree a good deal faster than I went up, collecting a few gouges in the process and feeling the orphan's claws as he slipped down my nakedness. By the time we reached the ground, the raccoon was anchored to the waistband of my shorts; he had made a number of bloody lines down both sides of my chest and across my stomach, and he had squirted a copious amount of loose feces all over me and himself. For good measure, and because he was now more frightened than ever, he bit my index finger, drawing more blood, when I reached down to pick him off my shorts.

In due course he was resettled against my chest, a rather messy bundle that required holding with two hands to prevent him slipping as I walked home at a brisk pace. Even so, he "squirted" twice during the journey and I was much relieved when he finally took up quarters in a cage.

During a much-needed shower, the orphan's name suggested itself to me: Squirt! When I was dry and after my various gouges and raccoon wounds had been attended to, I dressed myself properly and we examined our new ward, though for this inspection I put on a pair of stout leather gloves, holding him as Joan sponged off his personal mess and, with a new sponge dipped in a strong solution of potassium permanganate, cleaned off his wounded leg. On close examination we saw that Squirt had twin punctures in the upper part of his thigh; he had missed death by a narrow margin, somehow managing to wriggle out from between the bear's great teeth and to escape while his less fortunate siblings were eaten. During the afternoon of that same day, examining the area around the den tree once more, I found the remains of the female raccoon some two hundred yards away from the den tree, on the edge of a small swamp. Trails in the grass and reeds, and a large, flattened area where bits of bloody skin, fur, and bone lay on top of the disturbed undergrowth showed that the bear had carried the female raccoon to this place before eating her.

In forty-eight hours Squirt settled down, no longer growling and snapping when we tended his injury; by the next evening he showed himself willing to stretch out on Joan's lap while he fed from the bottle. By now he liked being stroked and brushed, and it was no longer necessary to shut him in his cage. When a few more days had passed, it was not unusual for me to enter the house and find Squirt draped across Joan's shoulders, like a live fur stole, obviously enjoying his contact with my wife and finding a good deal to interest him in the constantly changing view that he got as Joan busied herself within our dwelling.

It took the best part of three weeks for Squirt's wounds to heal, but many months were to pass before his pronounced limp became completely eradicated, a fact that caused us to care for him beyond the customary time and that almost led to his death some weeks after his rescue.

It was our practice to incinerate our garbage, using an empty forty-five-gallon drum, the bottom of which had been punctured a number of times to provide draft, the top having been removed with a cold chisel. This simple incinerator had a minor disadvantage—it lasted only about two years, the steel slowly breaking down under the assault of the flames. But when this happened, I would prepare another and cart the old one to the central garbage dump that the township provided. Not long after Squirt came to us, I had removed the lid from a drum but had not yet punctured the bottom. Other affairs intervened and inasmuch as the old burner was still serviceable, I left its replacement behind the barn, forgetting to turn it on its side.

One evening, I came home to find Joan in a sad mood. Squirt had not shown up all day; Joan looked for him as far away as the maple woods and called him repeatedly, at intervals, during the entire afternoon, without success. Changing my clothes, I went out and did some searching of my own, widening the area to a point where it encompassed a circle some two miles in diameter. I returned home at dark and we concluded that Squirt had wandered and that he would probably show up when he became hungry. That night, and all of the next day, Joan stood almost constant vigil; when I returned from work, she announced in tears that we might as well consider that Squirt had been killed by something, perhaps by "that horrible bear." More to comfort her than in hopes of finding the missing raccoon, I suggested that we take one last look, but as we prepared to search the maple woods it occurred to me that I hadn't checked the immediate surroundings of the farm, assuming that my wife had done so as soon as she noticed the raccoon's absence. It turned out that she hadn't thought of it, believing that Squirt would have responded to her voice if he had been hiding nearby. We discovered that the raccoon had, indeed, been close to home all the time that we had been looking for him!

Belonging to an inquisitive species, Squirt evidently had scaled the upright fuel drum, balanced on its edge, fallen inside it, and was unable to climb out again because of his gimpy leg. Hungry, thirsty, and exhausted from struggling to get out of the container, the young raccoon was in a bad way, too weak to call loudly.

As we walked through the back door of the barn after searching the building thoroughly, I heard a feeble scratching coming from the drum, looked in, and found the errant wanderer. Moments later he was in Joan's arms, and her tears were coursing anew and in greater spate, but this time motivated by joy. By the next evening Squirt was almost his old self again, and I was no longer being deservedly scolded by my wife for my carelessness.

On the evening following Tundra's pursuit of the wild bear, as I reviewed the way in which Squirt, Snuffles, and many other young wild animals had accepted Joan and me as foster parents, I concluded that though selfishness is frowned on by my own species, in the wild it is an extremely important part of survival, especially in the very young. By the same token, because of the demands of self, infant organisms, including those of the human species, are not at all motivated by love.

Later on, when the intellect begins to function on a conscious level, emotional attachments form, but until then, a mother's love for her baby is not reciprocated by the latter. Given food, comfort, and security, infant organisms of all species will respond as well to strangers as they will to their mothers, soon learning to trust the foster parents and to look to them for protection. These things did nothing to help me understand the spiritual kind of relationship between Joan and Snuffles that I had observed when returning with Tundra, but it occurred to me that perhaps the females of all species possess maternal attributes (an aura, if you will) that are easily detectable by infant organisms, but are hidden from the senses of adult males (I am not, of course, referring here to sexual signals that are intended to lead to mating and procreation). Even today I have not gone beyond this postulate; it may, or may not be the case, but there is no doubt in my mind that Joan emanated her own kind of maternal signals that were quickly picked up by the young animals that we cared for. They also responded to me, of course, but in a different way, and if the two of us happened to be present when one of our wards became frightened, it was to Joan that the little one ran for comfort and protection. In her absence, they chose me; but I was always second best in this regard.

Tundra clearly believed that he had won his spurs when he dared to charge the adult bear; he now set out to prove his newfound independence by intensifying his feud with Snuffles, who simply wanted to be left alone. The dog, exercising every hunting instinct that he had inherited, stalked the bear, ambushed him, and even resorted to trickery on occasion, pretending that he hadn't seen Snuffles, then suddenly wheeling around and charging him. The malamute was becoming a pest or, as even his doting mistress expressed it, "a nasty, aggressive little brute!"

Snuffles took to spending more and more of his between-meals time roosting in a tree like some huge, misshapen, black bird, while Joan, trying to prove to the dog that his jealousy was unfounded, took him on frequent walks.

One Saturday afternoon when I was busy writing and Snuffles was snoring in a tree near the barn, Joan called the dog and led him past the evaporator house and into the maples. The two had been gone no more than an hour when they returned, my wife's unusually hurried entrance telling me at once that something untoward had occurred. I was rising from the desk when Tundra dashed upstairs to greet me, his tail going like an aircraft propeller, his entire being radiating excitement and pleasure. Moments later Joan arrived, her face troubled and the spark of outrage in her eyes.

"You're going to have to do something about him! D'you know what that . . . that . . . that *beast* did when we were out? He chased a bear, that's what he did; a big one. It ran toward *me* and climbed a tree not more than a few yards from where I stood . . ."

I interrupted Joan's outburst, seeking to calm her and to get the story in a more coherent form, but my efforts were unavailing. Later, once more her calm self, she admitted that perhaps the bear had not been as close to her as she at first thought, but that it had been close enough, no more than fifty yards away when it had decided to tree. When it came to the animal's size, however, she stuck to her guns, maintaining that it was *big*. Perhaps it was the one that Tundra had chased a week or so earlier, when he was out with me?

It was entirely possible that the dog had found two wild bears within a ten-day period; these animals were relatively plentiful in our area and often came to the yard at night to pick up deadfall apples from several trees that grew in view of the house. We had one of those self-regulating, seven-hundred-watt lights that lit up the space between house and gate with almost the same intensity as a searchlight, but this did not bother the stolid bears. The animals did not interfere with us, or with our buildings, but came and went in peace as and when they happened to find themselves in our vicinity. They also spent a good deal of time feeding on the wild leeks within the maples, and we had seen as many as three on a single occasion; but every such meeting had, hitherto, been pacific, ending when the bear, or bears, ran into deep cover or, if surprised when we were close to them, escaped up a tree. Joan was aware that these big animals meant her no harm, but she thought—and quite rightly—that they could kill the dog or even do her accidental injury if she happened to be in their way. I agreed with Joan; something was most definitely going to have to be done about Tundra's bear hate. But what?

The next day I replaced a section of roofing that had been damaged by ice the previous winter and when I finished the job and was down on *terra firma* once more, I failed to remove the twenty-foot extension ladder when Joan called me in for a cup of tea. Other matters occupied me after that, and the ladder remained leaning against the roof.

It was when I was busy feeding Fang and Tame, and while Joan was preparing supper for Snuffles and Tundra, that I realized I had forgotten the ladder, but before I was ready to remove it the bear cub showed up. His appearance immediately triggered Tundra's enmity. Hitherto the dog had been following me, evidencing curiosity before the cages of weasel and shrew, but behaving himself otherwise. But when Snuffles arrived, the dog rushed at the cub before I could stop him, and the bear, finding himself near the ladder, began to climb it, bounding easily from one rung to the next until he scrambled onto the roof slope. He stopped when he reached the chimney and there, surveying his world from this new vantage, settled down, leaning his back against the bricks of the smokestack and propping his legs on the uphill slope of the roof, his supple body nicely fitting into the V made by the perpendicular chimney and the quarter-pitch angle of the rafters. At the bottom of the ladder a frustrated Tundra was left wondering where his enemy had disappeared to.

Snuffles evidenced a great liking for his perch. From that day onward he used it as his regular resting place whenever he happened to be in the vicinity of the house. After he was full grown, he still showed a preference for it, reaching it easily without the ladder by climbing up a tall elm that grew nearby and lowering himself onto the roof from an overhanging branch.

He supplied distinctive ornamentation at those times when he reclined against our chimney; and when he wasn't there, the piles of dung that he deposited on arrival or departure added what I thought was a distinctive touch to an otherwise undistinguished expanse of gray shingles. Joan demanded that the bear pats be removed at intervals, and I obliged; but I always left one mound. This allowed me to claim that our abode was rather unique, being, in all probability, the only one in North America to be so decorated.

My wife didn't agree. But I had my way because each time that she protested about my "filthy beast," I offered her the opportunity of climbing up on the roof so that she could do the job to her own satisfaction.

Eight

When the first proper snowfall of winter comes at night and spreads a glittering mantle over the land, and after the exhausted clouds have been sailed elsewhere by the wandering storm, the blue and white and brown and green hues of breaking day conspire to bewitch. The wilderness seems awed by the beauty that has been created; it falls silent while the red-orange sun edges into view behind the spires of the eastern trees, there to pause briefly, then to continue upward to become a glowing rind that backlights the far evergreens and dresses them in jeweled robes. The purples and pinks and mauves of predawn fade and are replaced by a cobalt infinity; the white that covers the understory no longer glistens as a single plane. Now the snowflakes are endowed with individual light; it is as though each six-rayed, exquisitely sculpted particle has been given a life of its own.

In mid-November, at dawn, I stood ankle deep in snow and contemplated the magnificent panorama that had been given to me during the night; and I knew that winter had arrived. Before this day, and despite the calendar and the daily broadcasts from the weather office, we had experienced the effects of a reluctant autumn, a season lingering past its appointed time and consequently in its dotage, drooling rain and sleet or spattering the landscape with undernourished snow that became quickly tired and melted into the earth. It was cold one day, unseasonably warm the next, and the sky remained wrapped in a gray shroud. But last night the storm had come; moaning and screaming, perhaps made petulant by the long delay, it pounced on the wilderness and loosed avalanches of swirling, ghosting white that cut visibility to no more than a few feet even under the yard light.

With Tundra I went out into the storm and we walked across the open ground and entered the shelter of the maples where the dead leaves had the power of flight and swooped and dived like fledging birds trying out their wings for the first time, mixing with the snow and often coming to stick themselves wet against our bodies. The dog was ecstatic. He ran in circles, rolled in the snow, returned to me to try and make me join him; and I did, now and then, but gave up quickly each time for I couldn't compete with him; this was his element. Afterward he stayed in the porch, on his chain, awake and watching the primordial display.

Joan was already asleep when I went to bed, and I was sorry, for I would have liked to talk; but as I lay there listening to the storm's elegy and believing that I would remain awake for a long time, the sound of all that splendid turbulence acted like a soporific. When I opened my eyes again, the first streaks of dawn haloed the trees outside the bedroom window and suffused the landscape with soft, fluorescent light. My wife continued to sleep, so I picked up my clothes and left the bedroom quietly, dressing downstairs.

Tundra heard me immediately and scraped at the door with a paw. I let him in, but made him sit while I finished dressing, not wanting him to gallop upstairs to disturb Joan. Moments later we went to greet the new day and to walk in the virgin snow, heading due east so that we could meet the sun when it rose clear of the trees and bathed our world with its gold. Soon we had company. The chickadees came first; some two dozen perky little birds festooned themselves on my person and called plaintively, asking for seeds; they ignored the dog as I held out a handful of sunflowers in which peanut hearts were mixed. As the chickadees each had a turn, the blue jays called, waiting for the seeds that they knew I would scatter for them on top of the snow.

We walked for almost two hours, Tundra and I, reading the messages written in the snow, the dog with his nose, I with my eyes, the silent language of winter that is as ancient as the first blizzard and as distinct as an illuminated text traced upon the finest vellum; it turned the wilderness into one great lexicon that recorded the passage of life and the visits of death; the tracks of a snowshoe hare emerging from under a white-mounded deadfall tree and disappearing again within the shelter of full-skirted cedars; wing marks, delicate traceries embracing three bright, crimsoned dots and superimposed upon the tiny footprints of a deer mouse, the whole describing the stoop of a saw-whet owl and the death of a fawn rodent; the sharply punctuated marks left by the feet of a deer dogged by the tireless paws of a wolf and his pack, a story that only the most determined "reader" could pursue to its ultimate conclusion; the meandering, patient footprints of a red fox seeking breakfast and meeting too late the neat tracks of a ruffed grouse that ended in a slight depression flanked by more wing marks; but they spelled escape, not death. These and many other stories caused me to fasten Tundra to his lead, for I did not want him to chase any of my forest friends.

So we walked, each being able to decipher the messages printed in the snow, the dog's sense of smell and my interpretive vision allowing us to understand what had transpired along our route since the clouds became deflated and the wind continued its turbulent journey, leaving tranquillity in its wake.

On our return, as we approached the barn, Legs materialized to view because she moved, starting to lope away when her keen nose and even more sensitive ears told her that Tundra was coming. It was as though a mound of snow had suddenly been gifted with life; the hare's white coat and her immobile crouch combined to make her invisible. Only a currantlike dot, her eye, and a small sprinkle of kicked snow to her rear indicated her presence. She did not stop, or turn toward us, as she would have done had I been alone, but loped swiftly to disappear into the shaggy, tangled mass of a dense cluster of hawthorns that had for years sought to choke a small apple orchard and which I had left because the slender, thorn-guarded trees offered such excellent shelter to folks like Legs.

Nearer the house, Tundra's sudden tension told me that some other form of life was in the vicinity. I searched carefully; movement attracted my eyes to a small, red-fawn shape that was busy harvesting seeds on the window feeder; it was Crazy, the white-footed female mouse who ought to have been eaten long ago, if only because she did not know the meaning of fear or caution, but who somehow continued to survive and to defy the odds stacked against her existence. She had first come to my notice three years earlier, in summer, when we were living in the gate cottage because I was remodeling the main house. The small, two-bedroom dwelling where once a hired hand and his family had lived was made out of cedar logs, each twelve inches long, that were plastered into the walls like bricks are mortared into place. The building was old and the mice had long ago made it theirs; nothing short of dynamite could have evicted them. I know; I tried . . .

With three live traps, I thought I was making some sort of an impact on the population, catching an average of seven every night, one at a time. When a trap clanged shut, it would wake me and I would rise, take the prison and its captive and walk to the barn, there to release the unwanted guest. Why the barn? Because I have a soft spot for white-footed mice; they are clean, interesting, and extraordinarily tame little animals. In the barn their chances of survival would be better and, in any event, there was nothing within the building that they could filch from me.

After getting by on very little sleep for two weeks, and while my nightly average of captives remained remarkably constant, I began to wonder about the efficacy of my plan, seriously considering the purchase of a certain kind of mousetrap that was able to capture six of the critters, each in its own little compartment. With three such multiples, plus my singles, I reasoned that I could imprison twenty-one mice per night and that I would not have to leave my bed in order to dispose of them so that the traps could be reset. Surely, I thought, a couple of nights would then be sufficient to clean out the remaining population? With these thoughts in mind I drifted into beatific sleep on the night that Crazy was to draw herself to my attention.

At 2 A.M. the familiar *clang* of the sprung trap awakened me. I rose, found a plump mouse in one of the three sets, and after a brief flashlight examination—during which I noted that this prisoner had lost the tip of her right ear at some point during her past and that she was more daring than most in that she continued to eat the peanut butter with which I had lured her into the trap—I walked to the barn, released the mouse, and returned to bed. An hour later another tinny bang awakened me. The earlier procedure was repeated and during the examination I noted that this captive also had lost the tip of its right ear; and it was also plump; and it was also a female. Too many coincidences, I decided: this had to be the same mouse. Didn't it? There was one way of making sure. I raided Joan's food lockup, a tin-lined steamer trunk that she used in favor of the built-in cupboards because it was critterproof, and I took from it one of those little bottles of vegetable dye that is used for coloring food; this one was shocking red. Into the liquid I dipped a pipe cleaner, shoved this between the mesh of the trap, and daubed red on the captive's backside. If, as I feared, the confounded mice were running back to the cottage as fast as I released them, I would have to alter my plans drastically. Should this be the case, I would know within another night or two when the marked captive obligingly reentered the trap.

I didn't have to wait "another night or two"; I only had to wait the half hour that it took me to clean up two piles of seeds that one of our unwanted guests had carefully stored underneath my pillow, a not unusual event resulting from the compulsive hoarding instincts shared by all members of the species *Peromyscus maniculatus*, who invariably seek out the softest, warmest places for this purpose. Drawers containing ladies' undergarments, being dark and cozy, have great attraction, as do blankets, housecoat pockets, winter boots that have felt liners, and, of course, bedclothes, especially when these have been warmed up by human bodies. I had just put out the light after removing the seeds and was preparing to seek new sleep when *clang* went another trap. I got up to look and found, sitting on the trip-plate and having a delightful time eating peanut butter, a lady mouse who was plump, was missing the tip of her right ear, and who sported the most gorgeously decorated bum in mouseland. My eviction system was not working. With the speed and accuracy of expertly thrown boomerangs, the mice I had been catching returned from the barn only slightly put out by being disturbed; I suspect they even got back home ahead of me! Disillusioned, I returned to bed, leaving the mouse where she was.

In the morning, after telling my wife of my discovery, I took another look at the captive and found her washing her face with dainty, white paws and not a bit concerned by either her capture or my presence. Joan called me to breakfast at that moment and during the meal she proved that she could outthink me by presenting me with a solution to our problem.

"You know, what we're doing here doesn't make sense. Why don't we just move into the house?"

Why not? The house was topsy-turvy and there was much yet to be done, but the kitchen and the bedroom were functional, and only the previous week I had installed the plumbing and the bathroom facilities; that the little room did not yet have a door was not that much of a problem. What *was* important was the fact that the house was *mouseproof*. We immediately moved in and thereby won the deer-mouse war—if one can attain victory by running away . . .

In any event, I released the decorated mouse and left her to her own devices while we wallowed in the luxury of food cupboards that actually contained food and in bedding that did not have to be inspected each night for little mounds of mixed seeds. Joan summed up our happiness in one laconic sentence: "It sure is nice to be able to put on a pair of panties without worrying about mice!"

The days passed swiftly, as busy days do, and I quite forgot about the one-and-a-half-eared mouse, but one morning, on looking at the window feeder, there she was, plump as ever, and still sporting a distinctive rump, though the dye had faded to a delicate shade of pink. I had seen white-footed mice abroad in daylight before, but they had always been scuttling along, en route somewhere, and trying to be as inconspicuous as possible. This one sat there in full view, munching. No wonder she was plump! That evening she was back, but this time she shared the feeding platform with Herself, who now and then reached for the mouse with a swift paw. The daring rodent simply scuttled out of the way, then, when the raccoon returned to her food, the little raider darted back to filch sustenance from under the jaws of death.

"That mouse is crazy!" I spoke aloud, but more to myself than to Joan. The name stuck. We saw a lot of Crazy as months turned to years and she continued to defy death as a matter of course. When Manx came, she shared his meals, sneaking up behind him while he was busy chewing, then darting in to pick up a morsel of meat and to disappear with it as swiftly as she arrived. She shared Snuffles's food, and Tundra's; and she continued to produce three or four litters of little mice every year. I don't know how long white-footed mice live in the wild but I suspect that their time on earth is usually limited to a year or so; that is why I am convinced that Crazy must have set what amounts to a mouse longevity record.

Joan was up and preparing breakfast for Tundra, who was now down to one small meal in the morning and a main meal in the evening. He had his matinal snack in the kitchen, but his supper, because it usually contained at least one large, raw bone that could grease up the rugs, was consumed outside. While he was eating that morning, before our own breakfast was ready, I went out again, going to the barn to check upon Snuffles. The bear was now rolled up in a tight ball inside a den made out of bales of straw, sleeping away the winter. Stooping low to listen at its entrance, the sound of rhythmic snoring assured me that he continued to "hibernate."

Leaving the barn, and closing and bolting the door behind me to prevent Tundra from getting in to disturb his rival, I decided that I no longer needed to check on the bear daily. I had been doing so for almost three weeks, ever since Snuffles was finally persuaded that bears really do go to sleep in the autumn and that they remain that way, unless rudely disturbed, until winter flees and spring makes life worth living again for the bruin tribe. Snuffles had been difficult to convince.

As the summer waned without noticeable change in Tundra's dislike of the cub, my wife and I worked hard either to keep the two apart or to continually try to get the dog to accept his companion, failing totally in the latter case and achieving only partial success in the former. By September, when Tundra was almost six months old and Snuffles was crowding his eighth month, the two had added inches and weight to the point where they could no longer be picked up easily; Joan could not lift either one of them, and though I continued to heft both in order to weigh them, I was always glad when the job was done.

Snuffles now wandered off into the wilderness with greater regularity and stayed away for longer periods at a time, while Tundra took to hunting groundhogs and was soon quite adept at the sport. With a mouthful of new teeth, he killed quickly, gripping his victims by the back of the neck and shaking his head violently sideways, snapping the vertebrae. This was not a pastime that either of us encouraged, but inasmuch as the groundhog population had increased to almost alarming proportions during the tenure of the octogenarian from whom we had bought the farm, we took comfort in the knowledge that the dog was thinning out the roly-poly chucks while engaging himself in the development of natural instincts. Then, too, because he didn't eat his kills but brought them home after he had played with them for a time, I usually was able to salvage the meat and the skins, preserving the edible portions in the freezer to use as critter food and tanning the skins, out of which Joan made warm mitts and other items of winter apparel.

By this time I had already sought to train Tundra as a pack dog, but he didn't take too kindly to this responsibility, and because I had not yet got around to buying a sled, the dog needed something other than long walks on which to blunt his vitality.

Snuffles, when not away in the bush, or reclining on our roof, loved to seek me out so as to wrestle, but after a number of such bouts I lost my appetite for the sport, not only because of the physical discomfort it entailed but also because it was hard on my clothing and necessitated the purchase of too many boxes of Band-Aids. He still managed to sneak up on me occasionally, but now I resorted to some of the defensive moves learned at great personal sacrifice from an army physical-training instructor who knew a lot of dirty tricks and delighted in using them. None of these were as easy to use on a bear as they had been on humans, but I scored enough times to cause Snuffles to think twice about ambushing me. And if Joan claimed that I was being mean to "poor old Snuffy," I countered by suggesting that she offer herself to the beast as his sparring partner. This usually silenced my wife's objections.

Life was not dull that autumn. Tundra, who also enjoyed his version of Kung Fu, would wait for me each evening and launch himself at my legs, the object of this being to fasten his teeth on one of my pant cuffs. If I saw him coming, all was well and we would roll about on the living room floor for some minutes of rough and tumble until I managed to secure a "scissors" with my legs around his waist, which soon had him banging the floor in surrender, metaphorically speaking. But if he got me when I was unprepared, just as I was taking a step forward, I would usually end up on my face, stretched full length, and sometimes butting the furniture. Now our "poor, bored pup" (my wife's words), would land on my back and nip with his incisors, being not the least bit careful as he took minute nips that produced brief but excruciating pain and left little purple marks in their wake. It may be supposed that I became extremely vigilant as a result of Tundra's "playful" activities; eventually I concluded that attack was the best method of defense. After that we developed a ritual. Every evening before supper we would wrestle in the living room during inclement weather or outside when the ground was dry and snowless. Indoors I would start the game by trying to slap him a couple of times and he would charge and grab my jeans, which I allowed him to do on the premise that if I was ready for him, he would not be able to tip me over. When he was busy killing one of my cuffs, I would twist around and straddle him and we would both end up on the floor until Joan spoke the magic word—supper!—at which point Tundra would streak into the kitchen and dance around my wife as she led him, food dish in hand, onto the porch. As soon as he started to eat, she would clip the twenty-foot-long chain to his collar and reenter the house, and we could socialize a little before sitting down to our own meal.

By the middle of October when the frost was getting heavier and the last autumnal hues were about to disappear, we began worrying about the bear's failure to den up. To encourage him to go to sleep I had constructed an elaborate, cozy nest inside the barn with bales of straw, making the walls two bales thick and the roof three bales thick. Leading into a chamber that was five feet square was a tunnel, rather like that built into Eskimo igloos, that was two feet high by two feet wide; beside this I placed a good mound of loose straw that was intended to block the entrance when Snuffles settled himself inside for the winter.

To accustom the bear to the den, I led him into it and fed him tidbits, things like marshmallows, peanut-butter-and-honey sandwiches, and other equally fattening junk. The cub was always content to snooze in his shelter for an hour or two, but he invariably emerged at the end of that time, pushing out the loose straw and waddling away into the forest. At first we thought that he might be looking for his own winter quarters, but since he turned up within a few hours and demanded an extra feed, we had to conclude that young bears are led into the bedroom by their mama, who thereafter sets the example by curling up and going to sleep. Well, I too could try *that* trick!

The date was November 2—will I ever forget it!—when, after a good meal apiece, Snuffles and I marched from the house to the barn accompanied by Tundra's yelps of rage and my wife's dulcet last messages. It was raining, but one could tell that the falling drops were about half a degree away from being ice.

Inside the barn, I sought to encourage the bear to enter his boudoir first, but he was reluctant, probably afraid that he would miss out on something. In face of his refusal, I dropped onto hands and knees and stuck my head into the narrow tunnel, a two-cell flashlight held between my teeth. I was nicely inside the confined passage when Snuffles elected to follow close on my heels, his clawed and clumsy paws urging me to greater speed when they scraped over my calf muscles. Reaching the lair, I squeezed myself against one wall, allowing as much room as possible for the rightful tenant, who waddled into the chamber but failed to show himself disposed to curl up. I put out the light, whereupon the cub stuck his runny nose into my face; I batted him on the head with my knuckles and growled. He didn't actually squeal, but he grunted softly and to my delight coiled into a ball and settled down. Have you ever tried to guess the passage of time while shut up in a totally darkened chamber? It isn't easy. After what I thought was a lapse of several hours, and because Snuffles was breathing hard and had not moved since he had been rapped on the bean, I shielded the light with my hand and peeked at the time. Half an hour had gone by since we entered the den! Even as I switched off the flashlight, Snuffles rolled over and put a paw on my leg, flexing his claws slightly, just to let me know that he was there and still conscious. Time dragged. I fell asleep.

When I woke up I felt awful; my neck hurt, so did the inside of my skull; the place smelled of stale bear breath laced with a soupçon of stale man breath; I itched in many places. But Snuffles was snoring! I peeked at my watch to find that it was 5:30 A.M. Snuffles continued to sound like a small motor. With painstaking stealth I edged toward the tunnel, reached it, and discovered that I could not go down it head first because my legs would have hit the bear. I spent several more minutes twisting around so I could sit upright and shove my legs into the passage. Eventually I slid into the barn, stopped to listen for the bear's snores, heard them, and stuffed armfuls of loose straw into the den mouth, packing it well. Outside the barn the rain had become ice and the grass and shrubs were stiff and glistening. It was cold, but I welcomed the feel of frost after my session in the fuggy den. I was quite certain of one thing: Snuffles would not be cold this winter.

That autumn saw the arrival at the farm of three new birds, two of which were the victims of hunters, the other rescued by me from the jaws of a large snapping turtle, though I suppose it was Tundra who really saved the life of the wood duck drake when he led me to the edge of the swamp where the drama was taking place.

It was in early September; Tundra and I were returning from a wilderness jaunt. Quite suddenly the dog began to pull hard against his lead, showing excitement and aiming toward the edge of a beaver pond where a profusion of cat-tails and marsh grasses grew. To humor him, and also being curious, I allowed myself to be pulled across the intervening two hundred yards, but it was not until we were more than halfway to the marsh that my ears picked up what Tundra had detected at the outset, a soft, continuous splashing noise that I at first mistook for the sounds made sometimes by a beaver when it is working on lodge construction.

I held Tundra on a short lead, to slow him down, and we advanced as quietly as possible; when we were within twenty yards of the marshy area, I knew that it was not a beaver at work; the splashes were too continuous, too repet-itive. By now I was extremely curious, but before seeking the origin of the noise, I secured Tundra to a nearby tree; he didn't like it, naturally, but I felt that his presence would limit my chances of solving the riddle.

After literally creeping through the underbrush and reeds, I saw the wood duck; it was almost immediately underneath a dead spruce tree that years earlier had fallen into the pond. The trunk was still somewhat solid, but the lower branches were tangled and rotting. It appeared as though the bird had somehow got one of its legs caught in the underwater ravel of dead sticks and that it had been there for some time, judging by the exhausted way in which it was thrashing with its wings in an effort to dislodge itself. The duck was so tired that it failed to exhibit fear when I stepped onto the log and walked slowly toward it; even as I bent down to reach for its body, it remained passive, but it dropped its neck and rested it against an outthrust branch. It was then that I saw the carapace of a big snapping turtle, a bumpy, pineapplelike shell, moss covered and some twenty inches long by about sixteen inches wide. I couldn't see the big reptile's head in the murky, roiled water, but it seemed that the turtle's beaklike mouth was fastened on one of the duck's legs. For some moments I was nonplused, not knowing how to free the unfortunate bird.

The struggle was taking place in the shoreline shallows, and the antagonists were at an impasse. Large as it was, the turtle could not use its greater weight in order to submerge, and drown, its victim; and it could not pull the struggling duck away from the tangled spruce branches. The bird, on the other hand, was firmly anchored by the thirty or more pounds of turtle; it could not free its leg from the tenacious hold. Soon it would become exhausted and would be dragged into the deep, where it would die.

Snapping turtles will pull down ducklings and goslings if given an opportunity to do so, this I knew; but I had never before found evidence to show that the reptiles also prey on adult birds. In the present case, unless I could free the duck, it was doomed.

Not able to reach the turtle's body from my place on the tree trunk, I thought I might manage to prod it with the tip of my hunting knife and cause it to let go, but after peering into the brown water without being able to see any of the reptile's vulnerable parts, I discarded the notion. As a last resort, I took hold of the duck by its neck, lifting it upward and hoping that I would not dislocate its vertebrae.

Holding the bird just below the head and trying not to squeeze any harder than was necessary to lift it, I pulled steadily, watching with alarm as the neck seemed to stretch to a point where I imagined I could hear its bones creak. As soon as the duck's body started to lift from the surface of the water, the turtle's weight made itself felt; a moment later the orange legs came into view. So did the turtle's ugly countenance; the bulging, dinosaurian eyes glared at me, but the evil, horny beak retained its hold on the duck's thigh. I kept pulling, believing that the reptile would soon let go; it didn't. When the mossy carapace was fully exposed, the turtle's neck stretched as tautly as the duck's. If I couldn't free the unfortunate drake quickly, it was going to die.

I went for my knife, only then realizing that I had instinctively taken hold of the bird with my right hand and that I could not now get at the weapon that was hanging on the right side of my body. Fumbling left-handed with the separate belt to which the sheath was attached, I tried to rotate it so I could grasp the weapon. The duck began gasping for breath; I was strangling it. My fingers touched the knife handle; I was going to decapitate the reptile. As though it had been able to read my mind, the turtle let go suddenly, falling with a great splash into the water and sculling down and away, seeking refuge in the depths of the pond.

Taking the nearly moribund drake to dry land I noted that the chelonian had evidently first secured a hold on the duck's foot, then inched its way up the leg, lacerating the leg as it traveled. The vicious beak had inflicted dreadful wounds on the bird's thigh; the blood was pumping out, but I managed to stop it by binding the leg tightly with a pocket handkerchief, rough and ready first aid that would have to do until I got home and was able to examine and treat the wounds properly. I was wearing a light bush jacket with a belted waist and I unzipped the top part, thrust the duck inside, zipped it up again, and freed Tundra. We started for the farm at a brisk pace.

Wood duck drakes are extremely beautiful birds. To see one in full breeding plumage during spring is to gaze in awe. It is as though nature, regretting the fleeting pause of the rainbow, had sought to incorporate into a single animal the riotous colors of the Arc of Iris, creating a bird whose kaleidoscopic, iridescent feathers would make it a fitting companion for the goddess herself.

At home I examined the duck thoroughly, finding that its foot and the lower part of the leg were not greatly lacerated; but the thigh had been badly mauled. It had several deep cuts and there was evidence of severe crushing and bruising; it was not surprising that the duck could not support himself on the leg. One wing, the left one, was also cut and scraped, but there were no broken bones, and I deduced that it had become injured when striking the spruce branches during the drake's efforts to escape. The bird would recover in due course, but it would not be able to migrate with its kind; we would have to keep him over the winter.

Even though his plumage was more subdued now that the breeding season was long past, he was still so magnificent that it was difficult to find a name that would do him justice. In the end, because he was a *drake*, we knighted him, dubbing him Sir Francis, but this was some days after he had taken up residence in the henhouse, settling in quickly and showing himself to be of calm disposition.

Wood ducks are so named because they nest in tree cavities, their feet being equipped with longer and sharper claws than those of other waterfowl. Bearing this characteristic in mind, I arranged a few wrist-thick branches in one corner of the building, keeping them low to the ground to allow him to climb up one-footed, without straining either wing or leg. I did not put in a tub of water, wanting to keep his injuries dry; instead I supplied him with a two-gallon poultry waterer.

I am not sure how long it was between the rescue of Sir Francis and the arrival of the snowy owl, but I believe two weeks had passed when I received a call from the Ontario Provincial Police office in Whitby, the same town in which Boo had landed. The policeman explained that this owl also had landed in a backyard, that it was evidently ill but showed no signs of injury. The bird was at the police station and would I like to collect it? On arrival I discovered that the owl had been housed in an unoccupied cell. There it sat on the back of a chair, yellow-eyed and inscrutable, the only honest-to-goodness, genuine jailbird that I have ever met. I soon agreed with the police examination; though the bird looked unwell, being listless and altogether too quiet, I could not find any injury; neither did it prove to have a temperature, which I discovered after getting from my animal emergency kit a rectal thermometer, the use of which caused more than one set of police eyebrows to rise almost to cap-band height.

At the farm, after subjecting the bird to a closer examination, I failed once again to find a reason for its obvious distress. When I offered it raw beef, it ate eagerly, and I was more puzzled than ever, until I noticed that it appeared to have some trouble swallowing. But when with Joan's help I forced its beak open and peered down its throat, my scrutiny did not reveal abnormalities. We named that owl Snowy; she was a big female, a magnificent specimen, except for whatever it was that ailed her.

Snowy owls *(Nyctea scandiaca)* usually inhabit the circumpolar, arctic regions of the world, wintering in those areas but probably making a short migration to points south of the twenty-four-hour, winter-darkness latitude. Feeding as they do on arctic rodents, particularly on lemmings, the owls tend to migrate to the southlands when their main prey animals undergo population cutbacks. These cycles in lemming numbers generally occur every four years, and as a result the owls are found in southerly latitudes during corresponding, low-population periods. An interesting side-effect of this bird's occasional invasion of warmer territories is the surprise that greets crows, or other large birds who seek to haze the owl, believing that it will not defend itself. The fact is that snowy owls hunt in daylight because of the demands imposed on them by the midnight sun of their home range. Although regional crows get a nasty shock when they seek to attack a snowy and find that their quarry can strike swiftly and fatally, they soon learn that the arctic visitor is a dangerous foe at any time, and so avoid it thereafter.

Since Sir Francis already occupied the henhouse and Snowy was too dangerous a companion for him, we housed the owl in our basement. There, sitting quietly on a perch, she fed well and survived for five days. On the morning of the sixth day Joan discovered her body, already stiff, lying on the floor. An autopsy revealed that the unfortunate owl had been hit in one nostril by a single shot-shell pellet, number-eight size, with a diameter of nine-hundredths of an inch. The tiny leaden sphere had penetrated her right nostril, leaving a mark that was only discernible after I had traced its trajectory in reverse. The shot lodged at the back of her throat, high up, where the wound was concealed by the curvature of the palate; she had slowly bled to death.

Snowy died a few days before the opening of the waterfowl hunting season, on the second evening of which, while walking home with Tundra as the banging guns spread death across the skies adjacent to the farm, we were nearly struck by a falling Canada goose.

Tundra was off his lead and walking a few yards ahead; neither one of us was paying attention to the skies immediately above and behind us. Thus, unnoticed, a big Canada gander was falling swiftly and at a steep angle, barely able to control his descent because he had been wounded. The first warning that I had of his imminent arrival came when my ears detected a sound similar to that which might be produced by rattling together a bunch of dried reed stems. Things happened so quickly from that moment on that though I ducked instinctively, alerted by the displacement of air immediately above my head, and even after I actually had seen the blur of the gander's body when it passed close to my face, I was still not sure what was happening until the bird plunged to the ground and fetched up against Tundra's back legs.

The dog was more than surprised! He leaped high, landed, ran off, checked himself and turned, dashing back, ready to attack the offender. The gander evidently had slowed up only just enough to allow it to land in a more or less upright position, but the collision with the dog's hind legs caused it to fall over on its side. It lay bloody and unmoving, its head up, its eyes fixed on the rapidly approaching dog. I yelled at Tundra and darted forward, getting to the goose before the malamute was able to attack. Picking up the wounded gander, I ran for home. Ten minutes later, for the third time that autumn, I became engaged in treating an injured bird.

The goose was exhausted, in pain, and suffering from shock, yet it was fortunate in that of nine number-four shotgun pellets (each a sphere thirteen-hundredths of an inch in diameter) only three had hit its left wing and none of these had hit bone. One pellet was lodged in the bird's left thigh, the other five had hit its breast, where, thanks to the heavy, insulating quality of the feathers, they had failed to penetrate deeply. Two of the wing shots had gone right through, the other had to be cut out; the breast shots were quickly removed with surgical forceps, but I had to incise the thigh wound before I could grasp the lead to remove it. The unfortunate goose bore the pain with relative calm, struggling only occasionally against the retaining bands of cloth that we had placed around its body, the ends fastened securely to my portable "operating table," a rectangular piece of plywood eighteen inches wide by twenty-four inches long over which I had secured a stainless-steel glazing plate from an old photographic drier, a piece of metal that was easily cleaned and kept in antiseptic order. From the sides of the wood a series of stainless-steel D rings served to secure the bindings used to immobilize injured animals during simple surgical procedures, though seriously hurt ones had to be taken to a veterinary clinic run by a friend of mine where I was allowed to use Jack's facilities, so as to undertake more complex repairs; when faced by really difficult wounds or injuries, Jack himself took over and I would act as his assistant. Strictly speaking, this Good Samaritan vet should have charged plenty for his services, for he wasn't a free agent, but he never did, and I shall always be grateful to him.

Honk, as we named the Canada goose, was to recover, but, like Sir Francis, he would not be able to join the southward-bound flocks until the following autumn. Goose and duck became companions in misfortune, sharing the henhouse amicably and, it seemed to us, glad of each other's company. Ducks and geese mate on a more or less permanent basis and become seriously upset if they lose their partners through hunting or because of predation. I have seen individuals of both species virtually commit suicide when they have circled over the guns after their mates have been shot down.

We had come through a busy spring and summer; now it seemed that we were going to finish up the year with only a marginal decrease in work, although I suppose that "marginal" element was not quite as small as might be supposed: Snuffles, after all, was fast asleep; his absence simplified our lives considerably.

When winter announced its presence with six inches of snow, and after Tundra and I returned from our walk, I felt that it was time for me to hand over to Joan most of the responsibilities imposed by our wild guests, leaving me free to devote myself to a considerable backlog of biological desk work that had accumulated since the spring. Unless I brought my notes up to date by the time the next vernal season came to fan new life into the wilderness, I knew that I would again forsake the desk for the out-of-doors.

Nine

While Snuffles was asleep that winter I engaged in a number of field trips in the hinterlands that lay north of our property, regions free of human habitation that were largely undisturbed and ideally suited for researching the habits and characteristics of those organisms that must remain active during the cold months. Mostly I went alone and stayed away for several days at a time, but on those occasions when I planned a twenty-four-hour journey, Joan and Tundra accompanied me, the dog pulling a light sled containing the extra camping equipment that was required for my wife's comfort, for I could not expect her to sleep under the stars, as I usually did, in order to travel light.

Joan enjoyed the wilderness as much as I, but she was not especially taken with the kind of long, arduous winter treks that I favored. I felt more at ease when alone, being free to travel silently and to roam wherever the objects of my study led me. For these reasons we only undertook two short outings between the beginning of December and the end of January, and when February arrived mild and sunny, the barometer indicating that this kindly spell would remain for at least a few days, I set off one Friday morning, early, with enough food to last me until the next Thursday.

The sun wasn't yet awake as I snowshoed away from the house accompanied by Tundra's laments, accoutered with a tumpline pack on my back and a haversack slung over one shoulder. My route lay due north, aiming for an area some twenty-five miles distant, where I expected to meet a bull moose, a four-year-old giant that had come to us when he was five days old after his mother had been killed by a truck as she attempted to cross a major highway. Snout, as we named him, had been free for two and a half years; I had last seen him the previous May, about a week after Snuffles was rescued. Now I was anxious to meet him again and continue with the task of plotting his movements, a study begun the previous winter and aimed at determining the territorial needs of these, the largest of all members of the deer family.

The calf weighed twenty-nine pounds when he was brought to us; when he left to find his own territory eighteen months later he must have weighed about 750 pounds. Last May, I had guessed his weight at about 900 pounds, which was close to the maximum that the breed attains in the more southerly parts of their range, an increase of girth that had been largely confined to Snout's hams and to a magnificent paunch that he had developed, a smooth and protuberant belly that proclaimed its owner's gastronomic inclinations and offered clear evidence that the territory he had selected was well able to support him in comfort.

He had been standing in the shallows of a beaver pond gathering the tender shoots of water lilies by submerging his entire head beneath the surface and probing for the young stems and leaves with his fleshy lips, all the while blowing bubbles and disturbing the bottom mud. Each time he secured a succulent mouthful he withdrew his dripping head and worked his jaws in concert with his long tongue so as to move the plants within reach of his molars. Then, chewing audibly and venting an occasional loud belch, he stood with eyes half shut and ears flicking this way and that as he routinely checked for danger. Later on, after he had filled the first compartment of his four-chambered stomach, he stilted his way out of the water and paused on the bank to scratch at his emerging, velvet-covered antlers with one back hoof before trotting away, going to lie down within the shelter of some nearby balsam firs to engage in the process of chewing the cud.

Within the concealment of a jumble of large rocks that decorated the brow of a hummock of land, lying prone, I watched him through the field glasses from a distance of about three hundred yards, where I remained for more than two hours, until he thrashed his way back onto all fours. Now, waiting until he voided the solid and liquid wastes that his kind almost invariably produce after such ruminative periods, I rose and began to walk toward him slowly, whistling a low, tuneless note that I had always used when approaching him. He had been about to wheel and charge into the protective forest when I got up, but now his ears stiffened as they picked up the whistle; he remained still, facing me until he evidently satisfied himself that the approaching figure was, indeed, the benefactor of his youth. A moment or two later he came to meet me.

I had brought him a present: half a dozen beetroots, a fondness for which he had acquired one morning when he was a gangly six-month-old. Joan had been about to prepare some beets, working on the counter in front of the open kitchen window, when the phone rang; she left the vegetables unguarded, and Snout, who had been standing some yards away watching my wife, ambled up to the window, stuck his head inside, and sampled the red delicacies. He loved them! Last May he had proved that he still enjoyed their sweet taste by eating half a dozen almost as quickly as

I could feed them to him. Now, in February, I was bringing him a new supply, fully expecting to meet with our gentle behemoth.

Raising the moose had presented few problems, perhaps because he was so young when he came to us; from the moment that he arrived, secured in the back of a pickup truck, he willingly settled down, accepting the bottle immediately. I remember that as he was sucking lustily on his second eight-ounce container, Joan, standing back and taking our photograph, exclaimed in delight: "Isn't he darling! And just look at his snout!"—which remark furnished his name. The resulting picture shows a placid, fawn-colored, leggy infant with his fleshy lips firmly stuck to the rubber nipple, ears at half mast, and big, black-velvet eyes staring benignly into space. Comparing this memory with the one I had obtained last spring, I found it difficult to conceive that the powerful animal I watched eating pond lilies was the same little calf that had leaned against one of our kitchen walls and sucked up the contents of three feeding bottles at one sitting, then bleated for more.

In this regard, it seemed that Snout had not changed; he had retained his enormous appetite, a not unnatural event for an animal that in the wild eats between forty and sixty pounds of vegetable matter in one day.

The one and only problem created by the moose occurred when he was six months old and big enough to be shot. That was in November—hunting time. He was too tame to be allowed loose to face the lurking sportsmen who, despite prominent NO HUNTING signs, tried to sneak into our property with monotonous and annoying regularity; but the law ordained that "pet" members of the deer family could not be kept in confinement during the murderous season, this legislation implying that all men who could afford a few dollars for a license were entitled to slay any animal for which they had a permit, regardless of whether or not somebody had gone to the trouble of rescuing it and sought only to conserve the species.

Joan and I respected all laws; we would obey this one, we decided, at least in principle. I did not force Snout into the barn; he followed me inside, evidently anxious to sample the handful of fresh beets that I carried as I entered the building. Moments later he was eating his favorite roots, standing placidly in the cattle section; he looked hot, so I opened a window that was three feet long and eighteen inches high and was located six feet from the floor; this, in a strictly legal sense, offered him a possible exit from the building if I should forget myself and close the door when I left. Assuming that he also would enjoy some good alfalfa hay when he had finished the beets, I opened up a bale and tossed it into one of the cattle mangers; then, on the premise that he would undoubtedly become thirsty after eating, I filled the water trough that stood in one corner of the barn. It was just as well I did those things, because I did, indeed, forget to leave the barn door open when I left. It seemed that for the next fourteen days Joan and I became troubled with lapses of memory! Each time we entered and left the barn, we invariably forgot to leave the door open so as to give Snout the right to go out and get himself shot. Two weeks later, a span coinciding with the end of the hunting season, our temporary amnesia disappeared as suddenly as it had manifested itself.

Entering the barn after the last gun had gone *bang*, I found the young moose standing near the doorway, seemingly anxious to ramble through the wilderness. Remembering what the law said, I left the door open on my way out; Snout followed me, accepted two beetroots from Joan, and headed for a tamarack swamp, still munching.

Moose appear ungainly and do not give the impression of

sleek swiftness common to other members of the deer family, yet these giants have been carefully designed for survival in an environment where cold becomes extreme and where snow may accumulate to depths of forty-eight inches and more. With their long legs, moose can travel through forty inches of snow without noticeable discomfort and can run at speeds of up to thirty-five miles an hour, two characteristics that frequently allow them to escape from their greatest natural enemies, the timber wolves. These predators can run at a top rate of forty miles an hour for short distances and at twenty-five miles an hour over long hauls, but they are greatly hampered by deep snow, which cuts their top speed by half and causes them to attain only about twelve miles an hour during a long chase.

Based upon my own observations, the sightings of other mammalogists, and numerous records kept by individuals as well as by government officials, it may here be stated that wolves kill only five moose, on the average, per one hundred chases (compared, for instance, with human-hunter success rates published by the Ontario government for the period from 1954 to 1962 that show that the lowest kill rate was 22.4 percent in 1962 and the highest 38.0 percent in 1954).*

More important than its speed and ability to run through deep snow is the moose's strength and aggressiveness. If the animal turns at bay, which it does more often than not, wolves will abandon the attack if they cannot induce the quarry to turn and run again within about fifteen minutes; the predators know that they are facing a dangerous antagonist that is capable of striking swiftly and with deadly aim with any of its hoofs, maiming and often killing outright any wolf incautious enough to come within range. In addition to

*_Some Facts About Predator Research and Management in Ontario_, Research Branch, Ontario Department of Lands and Forests (now the Ontario Department of Fish and Wildlife).

this, even when running with one or two wolves actually fastened to its body, an adult moose often will kill its hangers-on by smashing them against trees.

Undoubtedly wolves kill many moose, but it is equally certain that they go hungry with remarkable regularity and that moose, in turn, kill many wolves. For these reasons the moose and the wolf have for millennia existed side by side in the forests of North America and northern Europe, maintaining an equitable balance in the absence of men armed with guns.

In regions where moose and wolves are left undisturbed, available evidence indicates that the big deer live within a remarkably small individual range, perhaps roaming over an area consisting of little more than six square miles. Many tagged moose have been found still living in the area where they were earmarked six years earlier; this led me to believe that I would be able to make contact with Snout within forty-eight hours after leaving the farm on that February morning, his habits during the past thirty months confirming the statistics that I had collected. In his case, our moose had so far confined himself to a range of no more than some four square miles within a region where a few of his kind still survived, spread thinly over terrain that discouraged most human hunters. Snout's home base encompassed country in which three large beaver ponds were located, each separated by land that offered excellent winter feed in the form of second-growth aspen, stunted hazels, balsam firs, willows by the banks of the ponds, and wild cherry, all preferred autumnal and winter browse.

It was in this part of the wilderness that I found Snout the previous spring, at which season moose seek out tender water plants before the advent of summer causes them to eat herbs and the leaves of bushes and trees, occasionally getting down on their knees and grazing like some strange and ill-formed cow, the length of their front legs preventing them from cropping grass while standing on all fours.

I had seen Snout do this a number of times, an action that caused his pendulous bell to swing from side to side, actually scraping on the grass, an incongruous appendage that has long puzzled mammalogists without so far revealing a concrete reason for its presence. Both sexes sprout this dewlap, though it is more prominent under the chins of the bulls; some biologists suggest that it may help to protect the animal's throat from wolf attacks, others opine that it may serve as a symbol of status within the loosely knit hierarchical system adopted by the species; whatever the reason for it, the bell worn by an adult moose is certainly a distinctive appendage that adds to the spectacular appearance of this giant of the forest.

By full sunup that morning I had covered no more than two miles in as many hours, dawdling along, allowing myself to be entertained by the sights and sounds of the snow-covered wilderness and enjoying the fine, not unreasonably cold weather. Humping the heavy load while constantly detouring to examine some particular part of my surroundings soon caused the sweat to run and made me remove my heavy parka and secure it to the packsack; this gave me an excuse to roam around for a time free of encumbrances, the better to examine tracks and to commune briefly with some of the animals that had made them.

Taking up the load again, I ambled along quietly, content and deliberately aimless and allowing myself plenty of time to absorb the affairs of natural life and to think about their meaning. For many years I had been applying myself to the study of those animals that remain active during a northern winter, seeking to understand more fully the ways in which the cold affects them, causes them to alter their habits, and has, during the process of evolution, equipped them with physical characteristics that allow each species to better withstand the constant hazards of the frigid climate.

To gain at least some measure of insight into the problems of survival in regions where the mercury often plunges to thirty and forty degrees below zero and rarely rises above the freezing mark for months at a time, I believe it is first necessary to personally experience the bite of frost and to monitor one's own physical and mental reactions to this kind of stress; but because such experiments do not readily afford clear answers, research must be conducted on a more or less ongoing basis. That is to say, one or two experimental contacts with the cold are not able to supply answers to the many questions that must be asked before even marginal understanding is obtained. I soon learned what frost can do to the unprepared body, but there are so many variables attendant to exposure, and so many of them depend on personal physical and mental condition, that repeated tests must be conducted.

When I first encountered the intense cold of the boreal wilderness it seemed to me that nothing could be expected to survive under such harsh conditions; but later, after I had been given more time in which to observe the many animals and birds that actually thrive in those inclement habitats, I became deeply interested in the survival traits exhibited by those life forms that must endure the gelid winters of the northern hinterlands.

During that first day of my walk toward Snout's homeland I entertained myself by renewing the extent of my knowledge of this subject and by observing the animals of the wilderness, now and then pausing in both tasks so as to sit on a downed log to rest and to make notes. On several occasions that morning, and again in the afternoon, I unloaded my pack and spent an hour studying a particular animal or bird. At noon, stopping for lunch, I forgot hunger when a flock of ruffed grouse drew themselves to my attention.

The sturdy, woodsy-colored birds advertised their awareness of my presence by peeping softly to one another as they momentarily interrupted their feeding, but when I remained still they soon ignored me, concentrating on a clump of cranberry viburnum trees that, thin and pliant despite the cold, had bowed low to the ground under the weight of snow. The grouse were pecking at the glossy, carmine berries, tart fruits that when picked in autumn make excellent cranberry jelly; they were now frozen but offered themselves to the birds as a nice change from their more common diet of aspen buds.

Covering the snow were dozens of three-toed tracks, each showing the marks of the special little "teeth" that the grouse grow in autumn to enable them to walk on top of the deepest and fluffiest snow, the little combs, growing on both sides of each toe, acting very much like snowshoes. Here was a good example of adaptation to a northern climate.

The ruffed grouse *(Bonasa umbellus)* is found as far south as California and as far north as Alaska, the Yukon Territory, and the Northwest Territories. In the warm southlands the bird has no need of special snow-conquering devices; it does not therefore sprout the little projections on its toes. When, and in what way, did the grouse perfect the ability to grow snowshoes in the autumn and discard them again in the spring? I probably shall never find the answer to this question, but I keep on asking it. Of a surety, the bird has been making its living on our continent for a long, long time, for its fossil remains found in the province of Ontario have been dated at fifty thousand years. But that does not answer the question . . .

The grouse's dietary adaptability is not so difficult to

understand; it consumes all things edible according to season, motivated not so much by taste as by hunger and food availability; insects, the leavings of predators, seeds, fruits, and the tender leaves of young plants are all readily consumed from spring to autumn, after which season the grouse subsists quite happily on the terminal buds of such trees as aspen, birch, and hazel, taking frozen fruits when it finds them and not hesitating to feast upon the leavings of wolves or other predators after the latter have gone.

When the snow is deep and the temperature extreme, the birds dive into the white covering at sunset, shuffle their compact, pound-and-a-half bodies, and make for themselves an "igloo," dislodging enough bottom snow to kick up and so plug their entry hole. There, secure and warm, they sleep, yet they remain alert for enemies and will burst out of their shelters like so many rockets erupting from their silos.

After the grouse had eaten all the red berries and gradually worked their chickenlike way into the deeper bush, I sat basking in the weak sun and made some journal entries. I had just put away the notebook when movement to my left attracted attention. About ten yards from where I sat, a snowshoe hare was emerging from under the full skirts of a large balsam fir.

Here was another example of winter adaptation, reminding me not only of our own Legs, but also of the important role that ears play in the fight against the cold. Without the prominent sound scoops, the hare would quickly fall prey to the first predator that came its way; if those thin appendages were not in some way protected, they would become so seriously frozen that they would literally drop off. Creation has designed the ears of all winter animals so that they will not freeze, just as it has designed the ears of some tropical animals to furnish personal air-conditioning—the enormous, continually moving ears of the African elephant, for instance, are designed to fan cool air over the animal's body and are also well supplied with large veins, located just under the skin, that allow it to lose additional blood heat.

In cold regions the opposite must take place, the ears must *conserve* blood heat and for this reason each is first stiffened by a shaped form of nonvascular cartilage that is in turn covered by a thin layer of skin and tissue fed by small capillary veins; lastly, the whole is insulated with fur. Near the base of each ear, where lie the muscles used to move it as an animal listens continually to the sounds of its environment, the blood vessels are bigger and lie deeper within the tissue, being further protected by two kinds of dense fur: long, outer, guard hairs and soft, waterproof underfur, the same pelage that covers the remainder of the body.

Another characteristic shared by northern animals is that they are usually larger than their counterparts in more southerly regions, the premise here being that the larger the body, the slower the heat loss, just as a bath of hot water retains heat longer than a cup of the liquid at the same temperature.

Hares belonging to the species that I watched on that February day *(Lepus americanus)* are born with long, web-connected back toes that are sparsely furred in spring, summer, and early fall, but are heavily covered for winter protection; to give the animal better traction, it also grows a coarse mat of hairs on the soles of its feet. These, and the webbed toes that are spread during winter movement, allow the animal to walk, run, and leap on top of the snow and keep its footing even on glare ice. For added protection, the hare changes its gray-brown coat in the autumn, growing a new one that is snow white except for streaky, gray to light brown shadings on the back and the merest suggestion of dark hair framing its ears, the *tout ensemble* combining to make the animal almost totally invisible against its predominantly white background.

Fur, feathers, size, special attributes, diet, shelter, speed, vigilance: all these are custom designed by nature to assist the survival of those animals that remain to do active battle with the northern cold; then there are those, like the bears, raccoons, groundhogs, and others that beat the winter by going to sleep beneath the snow.

During early evening of the second day, I made camp in a place that I judged to be about the center of Snout's range, being guided to it by numerous moose tracks and by liberal piles of droppings and urine stains, the former having embedded themselves in the snow, the latter showing as yellow, pencil-round craters. I could not be sure that these marks had been made by our own moose, of course, but I was confident that it was Snout who had been here within the last two or three days.

That night, snug under a lean-to shelter made of poplar poles lined with heavy polyethylene plastic and further covered by a thick layer of evergreen boughs, I ate supper in front of a roaring fire and afterward drank hot coffee and made more notes. A half-moon emerged in a clear sky as I climbed into my down-filled sleeping bag. My last conscious action was to blow into the pint glass jar to extinguish the candle that was stuck to its base. Soon I was asleep.

At two o'clock I was awakened by the cold to find that the fire was down to a bed of coals and that the universe was ablaze with stars and further made brilliant by the green, pulsing light of the aurora borealis. The night was totally quiet; the coals snickled and crackled lightly, the sound of my breathing was easily audible, but the wilderness was as silent as outer space. I put new logs on the fire, waited for it to blaze anew, added some more fuel, and crawled back into the sleeping bag, there to listen to the stillness as I had done so often in so many other places while the night stalkers rested during the middle watches of darkness.

A gray jay awakened me just as the sun was creeping over the eastern tree line, a fluffed-out, gray, black, and white panhandler that was perched on top of my shelter cooing and whistling, its intent clear: it wanted to share breakfast with me. This was a scout, the first of several that lived in that area whose actions would be copied by its neighbors the moment that I offered food.

Within minutes of giving the first jay a piece of bread, two more arrived, one landing on the snow almost at my feet and the other taking station on the edge of my bivouac; when each had flitted away to hide its prize in some cranny of the forest, I made up the fire, cut off half a dozen slices of bacon from a one-pound slab, mixed bannock dough, and placed a potful of snow on the fire for coffee water. Before I was finished the three jays were back and I broke up the last few slices of bread I had brought with me and scattered the frozen crumbs around the campsite, letting the feathered gate-crashers help themselves.

I had done the washing up and was shaving with the help of a small camp mirror before the sun had traveled high enough to make itself felt. The day noises started to come in quick succession. The nuthatches became busy, calling huskily as they searched tree bark for dormant, hidden insects; the chickadees copied their neighbors, peeping as they busied themselves with the important business of food gathering; seven ravens arrived noisily, landing on treetops adjacent to the camp, there to sit and chuckle as they inspected every inch of snow around fire and lean-to in hopes of spotting some edible thing that they could steal.

When my whiskers were scraped off, I rinsed the razor, put it away, and washed my face with handfuls of snow, for there is no better aftershave "lotion"; dressing for the bush, I strapped on the snowshoes and donned the haversack in which the frozen beetroots, more like cannonballs by this time, reposed; lastly, I put in the satchel a bag containing mixed nuts, raisins, and raw, rolled oats, my snacks for the morning. I then went in search of Snout.

It took an hour to find him, the first half being spent searching for fresh tracks. Eventually, climbing a rocky rise and skirting a thick stand of white spruce, I found the place where a moose had spent the night, a characteristic large hollow in the snow located just inside the shelter of the evergreens and sprinkled with a number of stiff, red-brown hairs. Nearby, in the open, the moose had relieved itself before walking away in a northeasterly direction, its splayed hoof tracks clear and fresh in the snow. I followed.

For half an hour I snowshoed alongside the meanderingly haphazard trail that often led to some young poplar and ended temporarily in a proliferation of tramplings under the tree, the terminal branches of which had been chewed off; at one place, a broken aspen that was three inches thick at the butt told its own story; the moose had stood on its hind legs so as to reach the higher, more tender branchlets, leaning his chest against the trunk and pushing until the tree snapped, a few light-colored hairs that adhered to the pale green bark about six feet from the butt furnishing evidence of the big deer's pectoral shove; after the tree had fallen, the animal had browsed the top, eating almost all the branch tips.

I found Snout standing in a fairly open area where a number of young aspen and a carpeting of bushes grew. He was chewing the cud, but he had seen me, his big ears, erect and stationary, aimed in my direction, his bulbous nose raised and testing the breeze for scent. The field glasses brought him close to me across the intervening quarter of a mile, showing that he was in peak condition, fat, glossy, and clear-eyed. He was standing broadside on, but looking at me, his stance showing that he was alert and ready to run, but not yet sufficiently alarmed to escape at a full gallop. I began to whistle and to walk slowly toward him. At the sound, he turned his body and pawed at the ground with one front hoof, sending several cascades of snow backward; then he moved, coming to meet me with his head high and his long legs pacing casually, an unhurried gait that covered a lot of ground with each jackknife stride.

He advanced a good five yards for each one I traversed, a superbly coordinated, fluidly graceful animal endowed with a wild and rugged majesty that was as inspiring as it was intimidating. Against a background of pristine snow, green trees, and cerulean sky, his great body, each muscle seen to ripple smoothly as he moved, shone like burnished copper. Twin jets of milky steam emerged from his nostrils and trailed over his head; some of the exhaled moisture adhered to the edges of his ears and froze instantly, sugar-small particles of ice that trapped sunlight and glittered like minute diamonds.

I stopped, allowing Snout the opportunity to meet me on ground of his own choosing. He was the host, I the visitor. It was only right that he should be shown such a courtesy; he needed to get my scent properly, and he needed time in which to recall my voice, to bring to the fore the misty memories of his association with me. I was speaking quietly and calmly as he reached my environs and stopped twenty paces away, sniffing audibly and actually curling his thick upper lip as he did so, his great head towering above mine and his large, limpid eyes searching my countenance. I held up a frozen beetroot; he tested its faint odor, put his ears in the relaxed position, and strode forward, quickly closing the gap that separated us. He took the offering from the flat of my bare hand and his thick lips stroked my palm, the bristly hairs on their edges feeling like stubble on an unshaven face; engulfing the rock-hard root he mouthed it, and it clicked against his teeth, making a sound similar to that produced when two pool balls collide gently; he drooled. A moment later came a loud *crunch;* the frozen ball was smashed between his grinding molars and he began to chew noisily.

We socialized for about fifteen minutes, standing about three feet apart, each receiving the effluvium of the other's breath and body, odors that were comforting to each of us because, by exchanging them, we engaged in the kind of deep, personal contact that offers the security of friendship. Every human home is permeated by its own, singular redolence, a composite of the highly personal scents of its occupants that blends with the odors of the foods and material furnishings of the domicile; this is the primordial smell of the den, now become the fragrance of the home that we all instantly, but unconsciously, recognize and accept as a welcome. It does not intrude, yet it is unfailingly there and detectable. So it was with Snout and me that morning; our scents mingled and became one, the heady aromas of our environment—in this case our "home"—adding their own reassurance and allowing us to meet and to relate as trustingly as two members of the same family sharing the security of their own living room.

When all the beets had disappeared down Snout's long neck, he stood quietly, now and then snaking out his long tongue so as to lick the last taste off his lips; viscous spittle, pink tinted, sprayed at me when he wiped his mouth in that way, but the gelid little drops could not interfere with our communion; I hardly noticed them.

Left: Gut, the fledgling robin, rescued after a snake had coiled itself around the bird.

Below: Goggles, showing reason why she was named, as, head up, her eyes are still able to focus on author and camera.

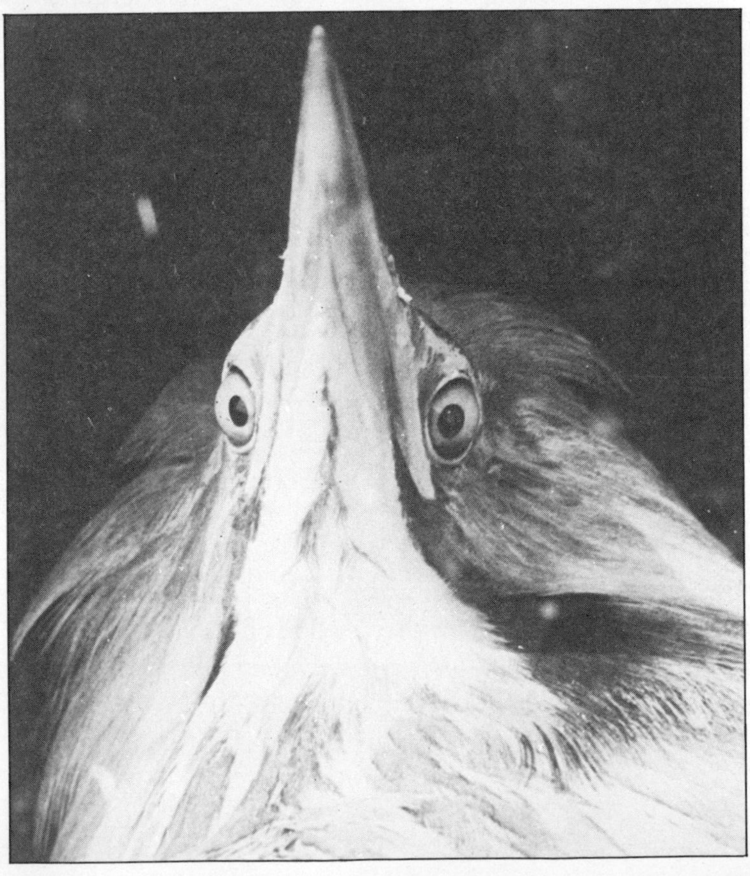

Snowy, the owl, who died from a shotgun pellet in the nostril.

Boo, the great horned owl, with feathers fluffed out to give the illusion of larger size, intended as an intimidating bluff.

Maggot, as an adult.

Red-shouldered hawk eyases, in the nest on author's property.

Left: *Sir Francis, a wood duck drake rescued from a snapping turtle that had grasped its foot.*

Right: *Blue-winged teal with a broken wing that author and veterinarian friend repaired by operating and inserting a steel pin.*

Below: *Honk, the Canada goose that almost hit the author when it was shot in flight by hunters and fell into one of the farm clearings. The bird was not badly injured and recovered by spring.*

One of two large flocks of chickadees that would feed from the hand (and out of the author's mouth, on occasion).

Killdeer chick found wandering on a busy road and rescued by author's wife, Joan.

Hummingbird that got caught in a burr and was freed by the author.

The spell ended when two ravens flew overhead and scolded us in their husky voices; Snout looked at the birds, snorted once, and took two steps to the rear. I started my snowshoe turn, moving carefully and slowly and refraining from glancing back at him until I had covered a good fifty yards. He stood where I had left him, staring at me, at ease, and once again working his jaws as he chewed the cud.

Two days later I was back at home and sitting relaxed before an open fire, tired, but peacefully satisfied. Outside it was evening; a few desultory snowflakes were being swirled against the windows.

At two o'clock in the morning of March 25, two days before Tundra's first birthday, we were awakened by the dog's staccato, high-pitched alert call, a furious and repetitive kind of latration accompanied by the rattle of his chain. The hubbub made us believe that some intruder was in the yard, incurring the big malamute's wrath. We were both befuddled by sleep and slow to respond—indeed, I was not inclined to respond at all, thinking that Legs, or some other one of our denizens had approached too near to the porch and thus aroused Tundra's displeasure. But under Joan's urgings and further prompted by our dog's continued outcries, I slipped into a terry-cloth robe, stuffed my feet into warm slippers, and padded downstairs, flashlight in hand.

Tundra was twenty feet from the porch doorway, at the fullest extent of his chain and struggling mightily to break it, or to at least pull the heavy retaining staple out of the wood. He was not responsive when I called and ordered him to be quiet and I was about to reprimand him and haul him back, when the sounds of violence being done inside the barn became audible above the dog's racket; at the same time, I heard the hoarse, groaning voice of our bear. Snuffles was awake and wanted out of his den!

The night was clear and full of stars. The snow of winter was almost entirely gone, with only scattered patches visible in low areas where drifts had mounded, solidified during the thaw-freeze cycles recurrent in March, and now lay stubborn and glistening under the refulgence of countless stars. It was chilly outside. I padded across to the barn door and opened it—and was nearly knocked down by Snuffles, evidently made almost frantic by his incarceration!

He emerged with a clatter and in a burst of pent-up exuberance, ran toward the house, turned around, and charged back to me, stopping just before collision with my legs and standing upright to put two paws on my shoulders and to peer into my face, snuffling. He could have done with some mouthwash! Yet I was glad to see him and I scratched his head and back, talking to him.

Joan emerged from the house, peered our way, and called out.

"Is that Snuffles?"

At the sound of my wife's voice the bear dropped back onto all fours and raced toward her. Tundra had other ideas! As the yearling cub neared the porch, the malamute began to snarl viciously, exposing all his fangs. The bear slowed, stopped, backed up.

Telling Joan to stay where she was and to leave the house door open, I ran to Tundra and slid my hand under his collar, pulling him clear of the entrance and holding him, despite his struggles, while Joan called Snuffles. The bear kept well clear of the angry malamute as he ran through the porch and into the house; Joan closed the door.

Seeking to calm Tundra, I remained with him for some moments, patting him and talking gently, but he wouldn't be placated. It was obvious that he now considered the bear an enemy and that he could not understand why the animal was allowed into the house. I felt too cold to stay and argue with him. In any event, I wanted to see Snuffles and to check him out after his long sleep and to note his particulars in my logbook; there would be time later to try and deal with the dog's antagonistic attitude, though I couldn't then see what was to be done about it. Ordering him to sit, I went inside in time to hear Joan calling from the kitchen. She wanted my help.

Entering the room I found Snuffles standing on his back legs in front of the food cupboard—the one he had crawled into last year. He already had opened it wide and was reaching inside with both front paws while Joan, who was no taller than the bear, tried to stop him from pulling out the marshmallows and other good things that were stacked on the shelves. The cub was nothing if not determined; and he was stronger than my wife, whom he ignored as he scrabbled with hands and claws until he hooked a package of marshmallows, then tried to get his mouth close to it.

Stepping between the two I rapped the bear on the nose with my knuckles, twice and ungently, growling as I did so. Snuffles reared away, dropped to all fours, and ran to hide in a corner, thereafter crawling under the table and whimpering in a manner that might have been pathetic if he had not been so large and so determined to get his own way. But the crying affected Joan immediately. She went to him, telling me as she did so that I should not have been so rough, an aside that I ignored. I closed the cupboard.

Minutes after Joan crawled partway under the table to fondle and coo at the spoiled beast, Snuffles recovered from his "rough handling" and emerged to view, pausing to eye me warily, and then, when I didn't move, to waddle toward the cupboard once again. He was being single-minded! I picked up a wooden chair, inverted it and did the lion-tamer bit, pushing the gangly cub away from the stove while Joan extricated herself from beneath the table. Snuffles backed away, tried to sneak around me, failed, and began to show signs of temper. He grunted and slapped at the floor with a forepaw, gouging the boards, whereupon I set the chair down, stepped close to him, and boxed his ears soundly. The varmint growled! But he retreated, then turned to go to Joan, who did not now look too happy to see her spoiled brat; she retreated also, going around the table.

It was evident that Snuffles meant no harm and had only growled in mild protest; it was also evident that the charade had to end quickly or Joan was going to develop a fear that he would detect, at which time he just might become dangerous. I spoke sternly to both of them: Joan was ordered to sit down and ignore the bear; Snuffles was growled at, then yelled at, and lastly advanced upon. He was intelligent enough to interpret the signs and to go and lie down in a corner where, just as if nothing untoward had occurred, he began to groom himself.

In all fairness to the bear, his behavior had been motivated by a natural desire to sample the sweet things that he remembered from last year, not because he was in any way vicious. In our anxiety to see him again, we allowed him into the house too soon, without permitting him to wake up properly.

My wife, the color back in her cheeks, rose from the table and announced that the cub was undoubtedly hungry and that she was going to give him some dog food. I sought to dissuade her, knowing that bears newly out of their winter den are still living off their autumnal fat because their systems have not been given the time to switch over from what might be termed self-consumption to the ingestion of food. Joan ignored my explanation, a determined look in her eyes.

Since I believe that all organisms have the right to learn by trial and error, I said no more, but went to station myself in front of the stove in case Snuffles, prompted by his sweet tooth rather than by hunger, was again gripped by the urge to sample some tidbits, which I felt sure was what he was *really* after.

Joan opened three cans of Tundra's favorite meat stew, poured their contents into a deep dish, and carried it to the bear. Snuffles, without rising, leaned forward, snuffled, stuck out his tongue, licked up a little of the juice, and returned to his grooming. I suggested to my wife that she now might like to offer him a few goodies, after which I'd take him outside and allow him to amuse himself at will. This time my advice was heeded.

Snuffles rose quickly when he heard the rattle of the marshmallow bag. He reached Joan's side before she had taken out the first sweetmeat, standing upright and trying to stuff his nose into the package; but Joan turned away and fished out a handful of the spongy goodies, then allowed him to take them from her palm, his prehensile lips picking them up as daintily as an elephant scoops a peanut off the floor of its enclosure with the grasping tips of its trunk. The bear ate five of the seven sweetmeats before losing interest in them; hesitating, snuffling wetly at Joan's hand for a moment or two, he allowed himself to drop back onto all fours before returning to the corner, there to resume his grooming, no doubt already starting to feel the itch that comes when an animal begins to molt.

With peace restored, I elected to wait until Snuffles had finished his beauty treatment before putting him outside, in the interval engaging myself with the coffee percolator. It was almost 3 A.M., but I knew that neither Joan nor I would be going to bed again now that the bear was awake and free to roam.

As the coffee bubbled on top of a low element, Joan and I talked about the bear's entry into the kitchen, for I wanted to know the details of his attempted raid on the food cupboard.

Snuffles, Joan said, had made straight for the cabinet as soon as he crossed from the living room into the kitchen, showing no hesitation: "As if he remembered exactly where the sweeties were kept."

Which was probably the case. He had, after all, remembered me and Joan and Tundra as soon as he emerged from the barn, so I saw no reason why he should not have remembered where his favorite tidbits were kept. Scent may have played some part in guiding him, but recollection, I felt sure, had been essentially responsible for his behavior.

The awakening of Snuffles seemed to signal the arousal of all those other sleepers who had spent the winter on our property. Scruffy showed up a few days after the bear; he looked more untidy than ever, but he had not lost his penchant for gathering food and seemed as determined as ever to risk his life in order to get it. I first saw him as he was cleaning out accumulated debris from one of his holes, after which chore he scurried toward the feeder, passing just beyond the reach of Tundra's chain and accelerating his pace only a little as the dog lunged at him. Seconds later he chased two chickadees from the seed station and began to stuff his cheeks.

Legs, who had been living sedately all winter, showed up one morning with a companion, a rangy, molting buck whom she was leading a merry dance in between the outbuildings, the poor fool not realizing that the more he chased her, the more coy she would become. The next day she had *two* suitors, and she really seemed to be enjoying the whirlwind courtship and the frequent altercations that developed between the hare Lotharios, who would pause in the pursuit of love in order to growl at one another, grunt, gnash their teeth, and then engage in a series of great leaps, simultaneously propelling themselves off the ground with their hind legs to heights of three and four feet while "boxing" each other with their front paws. As quickly as they started these bizarre bouts of fisticuffs, they would stop as though by mutual agreement in order to resume their amorous pursuit of Legs, who always crouched nearby watching the two boneheads fight for her affections. This behavior, I knew, would go on for at least two weeks, perhaps even a month, before Legs would at last make up her mind and choose one of the suitors.

Penny showed up soon after Scruffy, her coat already made ragged by the molt and by the demands of her system, which was having to nourish the embryos that obviously were developing within her. She was ravenous, demanding three feeds a day and searching every inch of the yard for edibles. Twice she came face to face with Snuffles, but the bear wisely allowed her the right of way. She didn't clash with Tundra because I kept him secured at this time of the year, a temporary measure that he was going to have to endure until I finished building for him a large, page-wire enclosure on the north side of the house, a place where he would be free to move about and from where he could enter the basement via a window and a ramp that I would make for him. In this way, he could see, but not touch, the small animals that came to visit and at the same time enter the house at will.

Herself turned up one evening, announcing her presence by standing up on the back feeder and placing both her grubby paws against the kitchen window, a stance that allowed us to note that she, too, would soon add to the population of North Star Farm. Exhibiting as good a memory as Snuffles had done, Big Mamma Coon, as we also called her, waited for Joan to slide back the window and then strolled in, stepping on the counter and waiting for her expected goodies and ignoring the fearful row that Tundra, outside on his chain, was making, for the dog had detected her presence.

When the maple trees began to weep sweet tears, Joan and I became busy, first preparing the evaporator house for the sap harvest, then tapping the trees, hanging sap pails, and, lastly, collecting and boiling the sap to convert it into syrup. For twenty-four days we worked from dawn to dark, snatching a few moments here and there to feed and care for our menagerie and feeling grateful that we did not have any very young animals on our hands. By the time that we had filled the last gallon can of syrup and cleaned up the utensils in the evaporator house, April had arrived and the woods were full of wild blooms and the bees had come, as had the ducks and the geese and most of the migratory songbirds.

One morning, as I went to deliver Penny's food, the sharp, two-syllable call of a hawk drew my gaze; planing some seventy-five feet over the outbuildings was Maggot, who now had company. Above him, but at least one hundred feet higher up, glided a large buteo, a female of the species, judging by her size. Maggot landed on the barn roof, fixed me with one piercing eye, and called once, asking as clearly as he knew how for some stewing beef. We didn't have any, but Joan had heard and seen her onetime enemy and she quickly emerged from the house bearing gifts that conveyed her forgiveness. The hawk hopped excitedly from one leg to the other, spread his wings several times, and peered fixedly at the supper steak that my wife had whipped out of the refrigerator and cut up quickly into hawk-size bites. She set the metal plate on the ground and stepped back as Maggot winged down from the roof and reached out a taloned foot, striking two pieces at once and settling himself to eat on the spot. His puzzled lady meanwhile continued to circle high, crying her plaintive call and no doubt wondering what her new mate was doing down there. This proved to be the last time that Maggot asked for a meal; soon afterward he and his mate set up housekeeping in a tall maple in the middle of the woods and there hatched three fluffy-white eyases.

We had half expected Boo to show up as well, but he didn't. He was still in the area and we had seen him on a number of occasions during the winter, once perched on a limb only about ten feet from the ground; we heard him call often enough—or at least, we presumed it was him because owls are territorial and the calls always came from within the immediate vicinity of the farm—but the bird had evidently found his rightful place in the wilderness and no longer had need of us.

Manx, Slip, and Slide remained absent, though I had not had time to look for the otters and hardly expected the lynx to turn up. Still, we were sad, even if somewhat relieved that the big cat was keeping to himself. But Babe stopped by—with her twins.

We found the deer in the maple woods one morning when we took Tundra for a walk; she was eating the sprouting leeks, separated by about one hundred yards from Pan and Pomona, who were now fully grown up and on their own. Babe was pregnant, her stomach quite bulgy; I guessed she was within three weeks of delivery. Pomona, not yet one year old, was not carrying a fawn and as a consequence looked as sleek and glossy as did her brother. Because of the dog, none of the deer came to us, but they didn't run away, continuing to crop and to pause now and then to gaze at us between swallows of pungent leek, the odor of which filled the environs.

As always seems to happen, spring, the most magical season of the year, slipped into the past surreptitiously. It seemed that one night we went to bed serenaded by the massed voices of the peepers only to wake up in the morning to find summer.

Joan decided that she would once more try to grow some vegetables and flowers, opining that now that Tundra was grown up and "so bloodthirsty" he might discourage the depredations of the groundhogs whose hordes, like the followers of Genghis Khan, had hitherto descended upon her tender victuals as soon as these showed proud and fresh-green above the earth. To this end, I emerged from the house one fine dawn and started up the roto-tiller, allowing my bones to be rattled out of their sockets as the infernal machine dragged me stumbling over the still-damp soil. By full sunup I had done about a quarter of the patch and was actually engrossed by my personal battle with the malignant, smelly, and rowdy machine, when I began to itch. Peering watery-eyed from out of the cloud of exhaust fumes that surrounded me, I became alarmingly aware that the entire female population of Ontario blackflies was having a convention immediately around my head, sucking up the carbon monoxide fumes like a junkie takes to heroin and spicing their heady "trip" with frequent sips of my blood. I didn't even stop to shut off the tiller; putting it in neutral, I ran for the house, removed my cap, covered myself in repellent, slapped gobs of the stuff on a white hard-hat I own, and returned to do battle with the land and with the eager millions of nasty little female flies that found me so totally irresistible, my only satisfaction lying in the knowledge that for every bite that I collected, ten blackfly ladies would become inexorably stuck to my hard-hat, trapped by the repellent. That's the only reason that I wear the plastic helmet, so I can count the victims later on and leave the cadavers until, by about October or November, the repellent dries and the corpses slowly disintegrate. It's my trophy hat!

That June after the garden was tilled, seeded, and badly protected by one lazy Alaskan malamute who suddenly seemed to have lost his zest for blood sports, I found myself resting on a large pile of rocks that the original homesteader of our acres had cleared from the land; there were many of these cairns, some, like the one on which I was sitting, held in place by a four-sided structure of cedar logs. I'd been squatting there for several minutes, my mind dwelling on the space requirements of different animals, when a faint squeaking intruded on my thoughts. At first I could not determine the direction from which the cry was coming, its lack of volume causing me to believe that it was issuing from some distance away; but listening more carefully, I at last realized that whatever creature was crying was doing so immediately underneath my seat.

Jumping down, I examined the rock pile and soon found the woodchuck doorway, located immediately underneath a large boulder. I listened at this entrance, heard the weak cry, but it sounded farther away. I listened at the top of the mound of granite boulders; it was closer to me from there. Curious, I began removing rocks, taking each one away carefully so as not to harm the squeaker, and working steadily for almost twenty minutes without making any discoveries. The squeaking had slowed, being now confined to intermittent outbursts that became louder in proportion to the number of rocks that I removed. Finally, picking up a boulder that must have weighed eighty pounds or more and was round, but somewhat flat, at the base, I found the caller, a baby groundhog, and his siblings. They were quite dead.

Judging from the little chuck's size and by the fact that his eyes were not yet open, it seemed that he was about three weeks old. His weak and emaciated condition, and the death of the other young rodents—there were five of them, their bodies still fresh, without odor—I had to presume that the mother had been killed by a predator and that her brood had starved to death, except for the lone survivor. Considering the groundhog war that we had been waging since we bought the farm, my first instincts were to summarily execute the moribund little animal. But then I felt pity for him and I gathered him up, put him inside my shirt, and carried him home.

Tundra, snoozing in the porch, became instantly businesslike as soon as he smelled my passenger; maybe he didn't want to exert himself out there in the fields chasing full-grown chucks under the hot sun, but if I had brought him a live one, well, he'd see what he could do about it! All he got was one close sniff and a bat on the end of his nose. He retired to sulk, flinging himself down with such petulant abandon that he shook the porch. Inside Joan asked the by now standard question: "What have you got *now*?" I told her.

My wife, as has been already said, was gentle: she hated bloodshed and she even killed mosquitoes and blackflies with regret, but when I told her that I had brought home "one of *those* brutes" she became instantly bloodthirsty. I had been unbuttoning my shirt as she pronounced the death sentence, but as soon as I produced the frail, pathetic little beast, Joan dissolved into nauseating maternalism. Taking him from me as though he were crafted in eggshells or Ming Dynasty china, she cuddled him and murmured incessant endearments. I went to prepare some formula in the faint hope that the orphan might yet survive; after all, he *had* outlasted all the others, so there must be some extra spark in him. Well . . . make it he did, and he was duly christened Charley inasmuch as he was a wood*chuck* . . .

Charley lingered at the portals of death for three days, but then, as though injected with the very elixir of life, he suddenly perked right up and moved his hitherto unmovable bowels, producing a number of oblong, yellowish pellets that melted to the touch, as Joan discovered when she picked one of them up in the mistaken belief that I had dropped some extraneous object into the woodchuck's shelter.

Thereafter the small groundhog surged forward in bursts of appetite and cascades of firm, this time green-hued, pellets. He soon acquired a rotund shape that was nicely set off by what my wife referred to as "a cute little potbelly," becoming as tame as a puppy during the interval and following Joan all over the house. His eyes had opened two days after I found him, signifying that he was now four weeks old (more or less); but his walking remained unsteady for the first week of perambulation, a fact occasioned by his infantility rather than by his period of starvation. He learned to climb on the toilet seat in order to try to get his own water, and this neat trick caused my wife to order me to remember to put the cover on the seat every time, which was something that I often forgot to do—until Charley almost drowned.

We were outside at the time, the woodchuck was alone within the house, theoretically supposed to be amusing himself. It may be that he really *was* thirsty, though I doubted that because his kind get most, if not all, the water they require from the vegetable matter that they eat; in Charley's case, he obtained plenty of liquid from the formula that he ingested in ever-increasing amounts. At any rate, he climbed on the seat, leaned too far over, and dived into the water, the slippery porcelain preventing him from getting out again. After what was probably half an hour, he was pretty well tuckered out, but Joan arrived in time to rescue him.

The month of June turned out to be Orphan Time at the farm. In quick succession came five very young gray squirrels, one fledgling robin, and one tiny field vole; caring for these and for the rest of our ménage kept us busier than we liked to be and caused Joan to exclaim plaintively one day, "We might as well be running a darned zoo!"

This remark led to a discussion; at its initial conclusion we were forced to agree that we were, indeed, running "a darned zoo." But yet, giving the matter more thought, we changed our minds: it wasn't a zoo! Or was it? More debate ensued. It was. No it wasn't. Maybe it was . . . for over an hour we dithered back and forth until Joan, her sense of humor restored, burst out laughing, rose from her seat, and stalked chuckling toward the kitchen. Pausing on the threshold, she turned to have the last word.

"O.K. Let's say that it's a zoo that isn't!"

The gray squirrels were brought to us by Billy, the lad who found Penny, but this time it turned out that the boy's father had cut down a tall cedar without realizing that it held the brood. The dutiful mother stayed with her helpless young while the chain saw bit into the trunk; she remained outside of the woven, basketlike nest that was suspended from several small branches, and was killed when the tree crashed to earth. But the five young ones were unharmed. They were fully furred and still blind and therefore not yet five weeks old.

The litter contained three grays, one gray and black, and one pure black, this being a characteristic of *Sciurus carolinensis*, a species that is predominately gray, but which often produces all-black offspring. Three of the young were males, two females, including the all-black one, the smallest of the brood. We didn't name them at that time, being too busy; later, after they reached what amounts to teenage in the squirrel world, the four larger animals left us to pursue their own way in life, but the smallest, the black, stayed on and became one of Joan's "darlings"; she named her Small.

The robin was discovered by Tundra, who inadvertently saved its life when he led me to some tall grass, poked into it with his nose, and revealed for me a large garter snake tightly curled around a fat fledgling. This was a characteristic of garter snakes that had been concealed from me until that moment; it reminded me of the large constrictor snakes found in the jungles of South America and Africa. I had seen many a rat snake constrict its victims, but hitherto I had thought that grass snakes existed on a steady diet of frogs and toads, caught by mouth, with a few of the larger insects thrown in as appetizers. In this particular case, the snake had obviously bitten off more than it could chew, if I may be allowed a very apt cliché! Despite the reptile's ability to unhinge its jaws so as to swallow large frogs and toads, there was no way that it could have ingested the big robin. In the end, it was not given the chance to try to do so, for I took hold of its head, intending to free the gasping bird, and the snake uncoiled of its own accord. The last I saw of it was when it found a crack between some rocks and scurried to safety as one of Tundra's big paws was batting at it.

The robin was not in good shape, but from what I could discover after palpation, none of its bones had been broken. After twenty-four hours of rest during which it built up a terrific appetite, it recovered sufficiently to stalk back and forth along a perch and to scream continually for food, all the while refusing to fly. Such were that bird's demands that I named him Gut, the appellation occurring to me very early one morning when, garbed in a bathrobe, I plied a shovel in order to find some worms for the avian glutton, who had awakened us both and would not be quiet. Unaware that Joan was sneaking up on me with a camera, I searched for the squirmy and slimy annelids diligently but with meager results. How often, dear lord of creation, had I seen dozens and dozens of fat, healthy earthworms gamboling on top of this particular patch of soil? Where had the creatures gone to this morning? After fifteen minutes of digging I found four midgets, as thick around as a piece of spaghetti, three-inch puny beasts that were probably too undernourished to dig their way to Australia as their brethren obviously had done. Gut engulfed them the way a gourmet polishes off dainty canapés, then screamed for more. The confounded bird was *insatiable*! Crawling around searching for bugs is not my most favorite occupation, but since the worms had migrated *en masse*, that's what I had to do, jar in one hand and tweezers in the other; my bag for an hour's safari was: three ants, one half-inch inchworm, obviously only half-grown, a green stinkbug that I allowed to go free, two houseflies (squashed), seven unidentified grubs, each three-eighths of an inch long, and, my great prize, nine larger tiger moths congregated on the porch screen—which was where I should have looked in the first place. This haul sated Gut temporarily.

The vole was found by me, unaided, while I was crawling along the ground looking for bugs. It was a tiny mite of a thing, fully furred, eyes open, but suffering from something that was to remain undiagnosed; it just sat hunched up, unmoving, not even when I picked it up and held it in my open palm, a blunt-nosed, mouselike rodent. We didn't keep him long, though, and he was as easy to please as Gut was difficult, accepting his diet of mixed seeds, eating slowly, but continually, while growing at a goodly rate.

Whatever it was that ailed him disappeared of its own accord, a fact that caused me to wonder whether he had been bitten by a shrew but had managed to get away.

By the time that Gut had finally learned to fly and, more important, to catch his own breakfast, the vole had gone to do what his kind are supposed to do down under the earth, and all the squirrels except Small had taken residence in the forests adjacent to the farm buildings; summer was getting long in the tooth, and affairs at the farm had slowed to a tempo that allowed us to have some spare time, more or less. Joan, who was dogged by Charley wherever she went indoors, had once more given up on her vegetable garden, but now, contrary to her previous outrage, she viewed the fat groundhogs with near equanimity and would not hear of their execution, her eyes always wandering to Charley whenever the subject was raised.

In July I found time to wander far and wide, to visit with Snout, and to socialize with Slip and Slide, now living together again at the lake and caring for a brood of little otters. Spike came to visit often and was as gentle as ever, though he did rather lord it over Snuffles and Tundra, seeming to know that both these rambunctious creatures would always fall back in his presence.

The feud between Tundra and Snuffles had grown worse, but, as usual, it was the dog that did all the warring and the bear that sought peace on the roof or up a handy tree. By this stage, the pattern had been firmly established: Tundra was a bear hater.

Both animals were large, healthy, and vigorous, each putting on the pounds as though in competition with one another, which I suppose they actually were. Tundra at this time weighed more than eighty pounds and the bear must have scaled at least one hundred, but whereas the dog had attained full height and length, though was yet developing musculature, the bear was less than half-grown, showing promise of being an exceptionally large member of his species.

Several weeks passed, the tenor of life at the farm-cum-zoo remained peaceable except for the dog-bear thing, but we had gotten used to it by now, and since no blood was ever spilled, we ceased to worry. I began a new book, Joan was taking a correspondence course in horticulture, perhaps thinking that she would learn some secret that would allow her to grow things unpalatable to groundhogs. *This was the life*, we thought; quiet, pleasant, free of major responsibilities—and then Tundra cornered Herself on the edge of a nearby swamp and started something that I finished at personal cost.

The dog's attitude toward the raccoon went beyond the hunter's natural predisposition to pursue prey; it was more akin to the antagonism that he displayed toward Snuffles. He had demonstrated his active dislike of Herself from the moment that he had first seen her, but the big raccoon, wise to the ways of the predators, had always been careful to approach the house only at those times when she knew that he was either secured by his chain or safely indoors. I repeatedly had observed her check to make sure that Tundra was not at liberty. She would approach unseen through the heavy timberlands, advance up to the edge of the road, and, just beyond the farm gate, suddenly appear in the same tree from where she had watched me on the day that she first drew herself to my attention. From her perch twenty or more feet from the ground she would inspect the property with eyes, ears, and nose. If Tundra was chained, she would begin to descend the tree, but slowly, pausing eventually to allow the dog to see her and waiting for him to lunge, as he always did, and be brought up short by the chain. When he demonstrated that he could not attack her, she would descend openly and almost arrogantly, stop momentarily on the ground, then move forward, taking a well-used trail through the grass that carried her close to the various trees that grew between gate and back feeder. If Tundra was loose, she would climb to the topmost branches of the tree, there to snooze until she was sure that it was safe to come down; but if the dog was inside the house she would descend moments after inspecting the yard. Inasmuch as Tundra, from within the building, would become aware of her arrival and, by his behavior, warn us of her coming, we did not find it unusual that the raccoon could also detect him when he was indoors—even if we frequently marveled at the acuteness of animal senses, and envied them!

One morning during early August I was outside putting the finishing touches on Tundra's 1,250-square-foot enclosure when I heard him lunge at the chain; silence followed the first rattle-crash of his charge, and I presumed that he had lost interest in whatever it was that had motivated him. Moments later I heard the deep and angry growls of a raccoon, the direction and volume of the noise telling me at once that the snarling animal was located somewhere near the entrance gate. At the same moment, Joan came to the side door that overlooked the pen area and told me that Tundra had got loose and had streaked for the swamp located on the other side of our road, opposite the gate. Interpreting the circumstantial evidence as I started to run toward the growls, I knew that the dog had finally caught the raccoon in the open.

The continual growling led me directly to the battleground. As I emerged from the trees onto the edge of the swamp, Tundra was thigh deep in the water, leaning back and shaking Herself, his teeth fastened on the back of her neck. I yelled at him, ordering him to "drop it," but he was too aroused to hear my instruction. I waded into the muck and grabbed him by the scruff, yelling more loudly. He released the raccoon.

She dropped, turned around to face us, and adopted the fighting crouch, head pulled well into the shoulders, mouth agape and full of gleaming fangs; she didn't growl now, but snarled malevolently and continuously. Tundra was again carried away with his passion and sought to charge her anew; struggling to hold him by the leather collar, I unwittingly moved one leg closer to Herself. She struck at me, clamping her teeth on my ankle, punching right through the leather of the boot. Joan arrived at that moment. She carried Tundra's lead, which she fastened around his collar, taking a few extra moments to do so because she couldn't find the D ring to which the snap-on fastener of the lead should be attached—Tundra had gotten loose when his lunge pulled the D out of the leather.

I remained absolutely still, feeling the teeth in my ankle. When Joan secured the lead underneath the collar and began to pull Tundra way, I let go of him, turned to Herself, and spoke soothingly, uttering nonsense, but seeking to reassure her by doing so in a calm voice. She let go and started retreating into the swamp, not turning her back until she was actually swimming and a good thirty feet away. She headed for a dead tree, climbed it, and settled down right at its very top. The crisis was over; the raccoon was not badly injured. At home I pulled off boot and sock to find that only three of her fangs had entered my skin. The punctures were not deep.

Two days later Herself came back, entered the house through the window, and allowed us to see that her thick neck fur, designed to protect her from the kind of attack delivered by Tundra, had prevented the dog's teeth from doing serious damage.

As for Tundra, the day after the fracas he was moved into his pen, there to wait until the local saddle maker had fashioned for me a collar, made from a hame strap, into which he secured a D ring that would have stopped the charge of a bull bison.

Ten

A few days after Tundra' s fight with Herself we were faced with a problem of a different kind when Small, the melanistic gray squirrel, unwittingly decorated with soot the interior of our house, climbing down the chimney in our absence and leaving hundreds of black tracks over the furniture, the rugs, and even on the walls.

Small's brothers and sisters were by now well settled on land of their own choosing, each pursuing the solitary life of its kind within the forested sections of our property, but coming regularly to the seed stations. Unlike her siblings, who had successfully made the transition from dependence to independence, Small did not show herself in any way disposed to give up the comforts of life within a domestic habitat. Although she continued to occupy at night the den that I had made for the entire brood and later affixed high in a big maple, the black squirrel spent about half of her waking hours in the immediate environs of our house and the other half inside, more often than not perched on Joan's shoulder, a pleasant little companion to whom my wife was quite devoted.

At the same time, Charley, the groundhog, continued to live in our home; he and Small had become excellent friends, companionable to the point where they frequently shared a snack at the container that Joan kept full of mixed seeds. Charley, because he chomped up the husks as well as the kernels, usually got the greatest share, but the squirrel didn't seem to mind as she deftly removed the shells from the sunflowers that she picked out individually, these being her favorite food.

Before Small came down the chimney I already had remonstrated against the continued presence in the house of the two young animals, fearing that they would soon become too tame and unable to return to their proper environment, and at the same time showing concern for the fact that they tended to deposit their wastes wherever they happened to find themselves at the moment of truth. If this were not enough, the presence indoors of the squirrel and the groundhog meant that Tundra had to be kept outside, and although the dog didn't normally enjoy being in the house for more than a short time, he did expect to be let in whenever he felt in need of our company. But Joan refused to listen. It was not that she was suddenly becoming stubborn, however, but rather that she had allowed herself to form too deep an attachment for the two wild ones. She assiduously cleaned up after them whenever they sinned, and she sought to make it up to Tundra by taking him for frequent, if short, walks and by allowing him to remain indoors for as long as he wanted to at night, after Small had left for her nesting box and Charley had been settled inside his own den, a "bedroom" that I had made for him within a portable cage.

I didn't force the issue, despite my misgivings—probably because Joan's eyes filled with tears whenever the subject was raised and because, if truth be told, I also had become attached to the interesting and intelligent young mammals, both of whom would have confounded the many members of my own species who still believe that the so-called lower animals are unable to employ reason to help them plan their everyday affairs.

No attempt was made by either one of us to train Charley and Small, yet both of them learned quickly to accommodate themselves to our routines and showed an uncannily accurate sense of time. They knew to the minute when we rose in the morning, when we took our meals, and when we habitually refilled their seed container. In some cases they learned by observation, in others through trial and error. Charley, for instance, could never again be induced to enter the bathroom following his dive into the toilet, and after Joan inadvertently stepped on his tail one morning because he had stationed himself immediately under her chair while we were having breakfast, he was thereafter careful to avoid a repeat of the experience. He soon learned to recognize our voices and the sound of our footsteps, but if a stranger entered the house, he ran to seek refuge in his den.

Small was not afraid of people, yet she distinguished between us and strangers, whom she did not approach closely. Outside, she was inherently cautious in dealing with all other life forms, especially with Tundra and Snuffles, and during those times that Maggot and his mate circled overhead, emitting their high-pitched cries, she immediately took cover.

When Small wished to come inside the house she would leap onto a windowsill and tap on the glass, showing that from outside she was able to detect our presence in any one of the rooms. To test her intelligence and persistence, we sometimes moved from one room to another, pretending not to notice her window tapping; she would follow, hopping from sill to ground and back up on a new sill, tapping and peering at us and frequently grunting hoarsely, until we let her in.

We were not unduly surprised when, during our absence one day, the squirrel learned that she could gain entry to our home by climbing down the chimney, but her unorthodox entry did cause me to immediately cover the stack opening with wire mesh. The small black squirrel made quite a mess, and if Joan pointedly refrained from complaining as she cleaned it up, I nevertheless could tell that she was considerably put out.

Two days later my wife discovered Charley trying to dig a hole for himself in one corner of the kitchen, his large and efficient chisel teeth carving great chunks out of the skirting board. That did it! Joan was at last convinced that a human dwelling was not suitable accommodation for a squirrel and a groundhog; she arrived at this conclusion unaided, but upon discovering that it was not easy to persuade her pets to forsake the good life in favor of the harsh realities of existence in the wild, she sought my assistance, keeping her counsel for an entire day and selecting the best psychological moment during which to solicit my aid—late at night, within our darkened bedroom, and while I was already half-asleep.

"You know, honey," she began sweetly, but in a voice loud enough to penetrate my semiconscious mind, "I've been thinking . . . you were right about Small and Charley. They'd be better off living free."

She said more, of course, but I didn't pay a great deal of attention to her words, being too sleepy. I believe I grunted agreement and promised to take a hand in the affair, but I can't entirely recall the nearly one-sided conversation. The upshot of this was that at breakfast next morning she asked me what I proposed to do about the problem, signifying in words and manner that, as the resident biologist, it was now entirely up to me to resolve the difficulties that I had not created. Aware of her attachment to Small and Charley, I knew that by taking what appeared to be an unreasonable stand, Joan was trying to minimize her upset, hoping that I would nibble at the bait she was dangling and turn the affair into one of our bantering arguments. I did. Even so, after some back-and-forth kidding interspersed with mock displays of anger, her eyes still filled with tears when I rose from the table and left to prepare a "safe house" for Charley behind the outbuildings, where there was one of those stone-filled cedar cribs that would offer immediate shelter to the groundhog.

To keep Tundra and Snuffles away from Charley's new residence, I once again made use of the electric fencer, connecting it to the chicken wire I had used to cordon off an area about the size of half a football field that would offer the groundhog plenty of feed. I could not protect him from above, of course, so we hoped that Maggot and his mate would stay clear of the neighborhood at least until Charley had settled himself in his new environment.

Joan accompanied me as I carried the groundhog outside and set him free. She had tears in her eyes at first, but they soon cleared away when she saw the obvious pleasure that Charley exhibited on finding himself loose on land that was well supplied with grasses, clover, and alfalfa. As we watched, the roly-poly little animal waddled around his new domain sampling the quality of the food it offered; but he forgot his hunger upon encountering the rock pile. Into it he went, appearing and disappearing among the rocks as he explored the tunnels and passages that they contained. Now Joan became wreathed in smiles.

Having settled one-half of the problem, we discussed the other half, Small. Since the squirrel was free and already the possessor of a suitable den, the main thing was to be firm with her when she sought to enter the house, a simple solution, but one that was difficult to put into execution because

Small well knew how to play on human heartstrings. I advised Joan to continue to allow her inside, but to gradually decrease the frequency of her visits. The best way to do this without falling for the squirrel's wiles was to spend more time away from the house; and inasmuch as my wife felt that she had been neglecting Tundra of late, becoming something of a servant to Small and Charley, she now resumed her after-breakfast walks with the dog.

We were still enjoying the heat of August at the time of these developments, the kind of weather that made it almost sinful to remain indoors longer than was strictly necessary. The next day was Sunday, and because I wanted to finish an article that had been interrupted by the construction of Charley's pen, I reluctantly stayed at home when Joan and Tundra left for their outing. Putting the dog on his lead just before her departure, my wife said that she was going to visit her "bee tree" in the cedar woods, an ancient, exceptionally large specimen that had a short main trunk some three feet in diameter and four secondary trunks that rose out of it at a point some six feet from the ground. Midway up the straightest and tallest of these thick ramifications lived a large colony of wild bees, the entrance to their hive being a round hole about half an inch in diameter located about twenty feet from the ground.

Joan was fascinated by these useful and clever insects. The previous year, exercising some of her abundant stock of patience and following the example of the late Karl von Frisch, the German authority on the habits of bees, my wife had become determined to test the time sense of the species, first attracting some of the insects to a feeding station that contained honey and then, with infinite care, marking with colored dots the upper part of their bodies. For this work she used refined shellac in which she mixed vegetable dye, filling four small pots with red, green, yellow, and blue paint. With a fine, sable paintbrush, she color-coded one dozen bees: four received one dot, each a different color; four were painted with two dots, and the last four sported three dots. Next she listed each bee in a notebook, referring to them by the initials of their colors: those that only carried one dot were R, G, Y, and B; the next group consisted of RG, GY, YB, and GR; and the last became RYG, GYR, RYB, and YYR.

This was the first study that Joan had undertaken entirely on her own, conceiving the idea, planning its development, and carrying it through to its eventual conclusion without any help, or advice, from me. And she did a terrific job! At first, with unconscious, yet typical, male chauvinism, I was skeptical, not really believing that she would be able to mark the bees, let alone obtain conclusive data from the experiment. But I kept my doubts to myself and left her alone, to win or lose unaided, believing that whatever the outcome, she would gain from the experience and she would derive personal satisfaction from the undertaking. Yet I watched her from afar, taking care to remain unseen; and I listened attentively every evening when she gave me a report of her day's findings.

She had noticed that a number of the bees from the colony came regularly to collect nectar and pollen from the lilacs that grew beside the house and she reasoned that if she stationed herself near these shrubby plants she would most likely be able to tempt her experimental subjects to feed from a little dish of honey that she put on a small table, on which she also arrayed her pots of color and three brushes. Anticipating a long wait, she placed a small stool beside the honey station, as she called it, and there she sat on the first morning of the experiment, commencing her vigil at precisely 8:10 A.M. She was delighted when the first bee landed on the edge of the honey container at 8:23;

delight turned to elation when, by 11:25 A.M. she had managed to mark one dozen bees out of several hundred that came to lap at the honey.

Pretending that I wanted to tidy up the machine shed, I had ensconced myself within this building and from there I watched Joan almost continuously. In no time at all she became oblivious to my presence. The first scout from the hive arrived, took its fill and left, but four minutes later nine more bees turned up, flying in and making a beeline (what else!) for the honeypot, denoting that they had been "told" of its presence and location by the complex little dance that the species have developed and by means of which they communicate with remarkable facility and accuracy.* Soon afterward Joan's honeypot was visited by a steady stream of workers, none of whom were disturbed by her presence; indeed, when she allowed one of her hands to remain close to the honey dish, some of the bees actually alighted on it and rested while they awaited their turn at the crowded container. It was easy for me to understand that Joan was seeking to accustom the insects to her proximity, so when she uncapped her color pots and picked up the first brush, I knew that she felt confident enough to try to color-code the first bee.

For two days she kept the bees supplied with unlimited honey during the hours of sunlight, then she restricted their feeding to five precisely timed periods each day, the first starting at 8:10 in the morning and the remainder being scheduled at different intervals throughout the rest of the sun hours, putting out just enough honey on each occasion to last the insects about fifteen minutes.

The bees were disoriented at first, coming repeatedly to

*See: Karl von Frisch, *The Dance Language and Orientation of Bees*. Cambridge, Mass.: Harvard University Press (1967).

the table and going away empty, but by the afternoon of the second day of the timetable, they had "set their clocks" and could now be counted on to arrive either immediately before or immediately after my wife emerged from the house carrying the dish of honey, but staying away during those hours when they knew that the honey would not be there for them. Joan had not doubted the findings of Professor von Frisch (nor those of other authorities), but she had wanted the satisfaction of confirming them for herself, of knowing the thrill that comes with personal discovery.

As an added bonus, she discovered that the bees, even the many that were not color-coded, but that came along anyway, knew and trusted her, alighting on her hands, clothing, and even on her head, without hostility. If she put a finger in front of one of the insects, it would climb on it, sometimes to wait a turn at the honey, at other times to sit and scrub its wings. As she so often said of them, "They're my *friends.*"

It was to be expected that her interest in the bees should continue long after she had abandoned the experiment. She visited the hive frequently, worried about the insects, often putting out honey for them. When Snuffles came to us, she asked me if I thought that he would do harm to the hive, but I reassured her by pointing out that despite the number of bears that inhabited our part of the country, some of whom often passed at least near the colony when they came to our apple trees, the bees had never been molested.

In any event, as soon as Joan and Tundra were on their way to the cedar woods that August Sunday, I settled in front of the typewriter and forced myself to concentrate on the half-written column, succeeding so well that I became oblivious to the passage of time. I was nicely into the swing of my theme and very much involved in it when a sudden hubbub downstairs intruded irritatingly. Accompanied by the crash of the closing door, Joan's voice, raised several octaves higher than normal, reached me, but if she was uttering words they were unintelligible. My annoyance vanished as I realized that something was altogether wrong. I hurried downstairs. Reaching the living room, I stopped, amazed and immediately concerned, by my wife's appearance.

Joan was crying and slapping at her hair and face with both hands and wriggling her body as though possessed by convulsions; her cheeks and forehead were inflamed, showing angry-red welts. Tundra was also convulsing. The dog was rolling his body over and over on the rug and he too batted at his face with his front paws. I stood nonplused for perhaps two seconds, noting that my wife was breathing with difficulty and was unable to speak, the sounds she was making emerging as moans and whines. I came out of my trance and moved toward her at the same moment that three bees disengaged themselves from her hair and flew to bump against one of the windows. I began to understand.

Slowly, hindered by shortness of breath and by the pain that she was feeling, Joan told me that Snuffles was responsible for her plight; he had followed her and Tundra, she said, and before she could stop him, had climbed the cedar and had started scratching and biting at the hive entrance. The insects boiled out of their nest and attacked indiscriminately, and she and Tundra had been forced to run from the scene, pursued and stung by a number of the angry workers. The dog had collected seven stings, but by the time that my wife finished explaining the problem, he had largely recovered, except that he now sported a swollen nose.

Joan wasted few words giving me her explanation, limiting her account to the bare facts because she was worried about the hive. Indeed, she was so concerned that she would not be satisfied until I agreed to go, then and there, and get the bear away from the bees, an assignment which filled me with considerable reluctance; had Joan not been so physically and emotionally upset, I would have refused point-blank; as it was, I bowed to her wishes. But I took precautions.

First I put on my boots, securing my pant cuffs over them with cord; next I donned a heavy jean jacket that zippered all the way to the throat and had a collar that could be turned up. After that I got my bush hat, fitted a mosquito head net around it, and put these on, making sure that the net fitted snuggly around the jacket collar. Lastly, I put on some work gloves and secured them under my sleeves with elastic bands.

The angry buzzing of at least ten thousand bees (an estimate—I didn't stop to count them!*) reached me moments before I entered the cedar woods, a sound unlike any I had heard before and one that did not cause me to relish the task that lay ahead. Arriving hesitantly at the heart of the trouble I was quickly enveloped by angry workers, hundreds and hundreds of them; they clustered so thickly on the face part of the net that I had to brush them off continually in order to see.

Snuffles clung to the trunk with difficulty, being forced to hang on with both back legs and one front paw. This limited his attack; as a result, it appeared that he had done little damage to the hive entrance, though I could not see well enough to make sure because of the bees that clustered around him and coated his body so thickly that it looked as though each hair had suddenly been given life of its own.

*A colony of domestic bees living in hives provided by apiarists contains between thirty thousand and fifty thousand insects; wild colonies are fewer in numbers, but contain between fifteen thousand and twenty thousand bees.

Yet he was all but impervious to the attacking workers; now and then he would pause in his scratching and biting of the wood so as to brush his paw over his face, mashing dozens of workers with each swipe, but he appeared to be determined to get at the sweet contents of the hive.

I yelled at him, growled, thumped against the tree trunk with my gloved fist. He ignored me. There was no power on earth that could entice me to climb that tree to pull him down! I looked around in hopes of finding a rock, or even a piece of branch, to throw at him; there was nothing, and I remembered that Joan had spent many an hour in this place and had "tidied it up" so as to create a parkland setting around her fascinating insects.

I tried to call the bear down once more, putting as much sternness in my voice as I could muster. To my amazement, Snuffles responded this time. He began to climb down, sliding rather like a fireman comes down the stationhouse pole. Once he stopped, as though he was about to change his mind and go back up, but I yelled again and he resumed his journey. Now I could see the hive entrance; it was hardly damaged, the wood being thick and strong and the hole offering little purchase for his claws. Located as it was on a branchless trunk, the hive entrance was virtually out of bounds to a marauding bear, a fact that probably accounted for the colony's continued success. Even an adult bear would find it impossible to broach an entry. Experienced bears probably wouldn't try to get at the honey, leaving it to inexperienced youngsters like Snuffles to find out the hard way that the hive was too well protected.

When our cub reached the ground I growled at him to show my displeasure and I reinforced the message vigorously when he turned to look back at the tree, as though he would climb it again; his rump offered such a tempting target that I booted it unceremoniously, at which he dashed away, heading toward the house. I followed, more slowly, but at a good trot, for now that I was the only intruder, the bees zeroed in on me in their thousands. But as I ran, more and more of the aroused workers fell away, and by the time I reached the house I had quite lost the buzzing escort.

Indoors, talking to Joan as I divested myself of the head net, I was jolted three times by a number of bees that had become entrapped in my clothing, but I felt it was a small price to pay for peace and I was rewarded by my wife's pleased smile when she learned that her beloved insects had not been eaten out of house and home by "that horrible bear."

It was not too long after the bee affair that we were visited by a colleague of mine and his wife on a Saturday afternoon, an effete couple whose knowledge of the outdoors was limited to the doings of the few insects that attended the ceremonial barbecues that they hosted in their backyard. They were what one might term "sort of friends" of mine, our association stemming from professional interests. As usual, they turned up equipped with individual pressurized cans of insect repellent that they began to spray at the environment the moment that they stepped out of their automobile. My colleague never visited me at the farm unless he wanted me to do something for him, so I prepared myself for a long and tedious session, knowing from experience that the object of this visit would not be disclosed until my guest considered that I had been suitably softened up by a circuitous preamble.

When preliminary greetings were disposed of in our living room, my acquaintance asked if he could talk to me privately about a matter that he felt sure would be of great interest to me. I suggested that we might sit outside. His wife, undoubtedly primed beforehand, said that she would rather sit inside and talk to Joan.

Before our party was split up, I offered drinks; sherry for the ladies, gin and tonic for my quasi-friend and me. Glasses in hand, we left the house and lowered ourselves into lounge chairs on the front lawn.

It was a pleasant, sunny afternoon during which the birds were active and in good voice; some of the lesser animals, such as Small and Beau Brummel, came occasionally to visit and to be rewarded by peanuts, a supply of which I carried as usual in one pocket. At first my guest and I chatted about nothing in particular, occasionally referring to mutual acquaintainces and exchanging news of their doings. We sipped gin and tonic, gossiped, and enjoyed the weather and the comfort of our chairs. But when the first drinks were finished and had been replenished, the matter that I was supposed to find irresistible began to surface. Before my visitor was halfway through what he fondly believed was a clever buildup that disguised the main object of his approach, I knew what he wanted, but knowing that it would be useless to try to halt his verbal flow now that he was well and truly launched in it, I remained silent, seeming to be listening, but my eyes and my attention occupied with the many interesting little developments occurring in our immediate environs.

My colleague was beginning to warm to his theme; I was busy watching some movement on the edge of the tree line, about two hundred yards away, a place that faced me, but was behind the visitor's chair. Some moments later the cause of the disturbance emerged to view; it was Snuffles, who directed his steps toward us. My interlocutor continued to talk unabated, using all of his considerable powers of persuasion to make the bait sound like a full-course meal; nodding occasionally as if I was devoted to every word that he was uttering, I kept my eyes on Snuffles, now halfway across the cleared space that separated us, but coming slowly and stopping often to sniff at interesting odors that he encountered on the way. Eventually the bear reached a spot some ten or twelve yards behind my guest, but inasmuch as the latter continued talking nonstop, I felt it would be useless to seek to interrupt him so that I could draw his attention to our bear.

It may have been the way that I gazed fixedly over his shoulder that alerted our visitor, or he may have heard some small sound made by Snuffles that caused him to turn around, gin glass in hand, to look over his shoulder at the moment when the cub was only six feet away.

My colleague was not exactly obese, but he was definitely overweight and made soft by self-indulgence and the effects of a sedentary kind of life; for these reasons I was utterly amazed when he reacted with a speed that was as unexpected as it was comical. The glass and its contents flew high into the air at the precise instant that my friend uttered a high and definitely girlish scream; at the same time, his arms and legs gyrated like the vanes of a windmill struck suddenly by the first puff of a tornado. So energetic were his movements that he could not at first rise. Wriggling frantically for some instants, like a great bug trying to climb up a sheet of glass, he at last became coordinated, found his feet, then dashed for the safety of the house, a record-breaking sprint that easily would have won for him a bronze medal in Olympic competition. Ironically, he was running without being pursued!

Snuffles moved slowly forward to lick up some of the splashed gin and tonic, made a face, then came to ask for a peanut while watching the fugitive in puzzlement, no doubt wondering why the stranger had hightailed it in such disorder. Reaching for a handful of peanuts, I became convulsed with laughter as I realized that the sudden and proximal confrontation with a real, live bear had filled my visitor with abject terror. The action lasted but seconds. Snuffles was just reaching for a peanut when the upstairs hallway window opened, and my shocked visitor's head poked out of it.

"There's a bear . . ."

The hysterical exclamation was cut short by the sight of Snuffles accepting a peanut from my fingers and daintily shelling it, while sitting himself down beside me.

I did my best to control my mirth, not wanting to hurt my colleague's feelings, but it was simply impossible! His initial behavior was funny enough, but when I realized that he had not stopped running until he had reached the highest part of our home, I broke right up. I couldn't help it!

Joan later told me that our visitor had burst into the living room in great agitation, making sure that he slammed the door shut behind himself. He had said not a word to his wife or mine, but ran through the room and scrambled upstairs as fast as he could go.

When our lady guest learned the facts and saw for herself through a window the slavering beast that was still accepting peanuts from me, the poor soul was taken with a fit of the vapors. The visit was cut short, its objective never quite surfacing. But the couple would not leave the safety of the house to walk to the gate and get into their automobile, even after Snuffles had gone to explore some other part of his world. Nothing would do but that I had to go get their car and drive it right up to the front door. Even so, neither of our guests would emerge from the house until I had gone out once more to open the offside front door, toward which they moved with great alacrity, each jostling the other in their anxiety to reach the shelter of their automobile. My colleague won. He led his wife all the way, at that moment showing little regard for chivalry and no doubt fervently believing in the equality of the sexes; it was clearly a case of *sauve qui peut*. Neither my colleague nor his lady ever visited us again.

In this way did Snuffles win himself back into Joan's good books after disgracing himself by seeking to eat wild honey.

It took me two full days to tone down the amusement that I felt every time I recalled the behavior of those two "city slickers." Time and again I would suddenly burst out laughing, even when engaged in serious affairs. Today, all those years later, I cannot help chuckling when I think of the incident and I still enjoy relating it to friends.

Joan shared my amusement; she frankly disliked my colleague and thought that it served him right. Nevertheless she did accuse me of allowing the affair to develop by not warning him of Snuffles's presence in time to give the man a chance to preserve his dignity. Not so! It never occurred to me that he would react in such panic, mostly because I was so used to having the bear around that I suppose I looked upon him much as a dog owner looks upon a pet that is known to be friendly and harmless. Nevertheless, funny though the incident was, it was worthy of serious consideration.

This occurred to me some weeks later, after I was told by other friends that my excitable colleague was heard to retell the story of his narrow escape "from a huge and vicious bear" that Lawrence kept on his property, the sort of exaggeration that commonly arises when those who are unacquainted with the wilderness and its inhabitants have had occasion to find themselves in close proximity to even the most inoffensive animals.

In quite recent times I was visited by a friend whose

knowledge of the wild was derived almost exclusively from Old World legends and New World hearsay. As he walked toward me on arrival, I was once again reaching for a peanut, this time to offer it to one of our present crop of friendly red squirrels. Nutsy is a female with a penchant for climbing up my pant leg in order to be closer to my hand when it emerges from a pocket clutching half a dozen peanuts. It was she who was heading in my direction as our friend directed his steps to where I was standing. Of course he saw the squirrel; and she saw him, for she stopped some distance away in order to appraise the situation.

"That thing," said my friend pointing to Nutsy, "is the most *vicious* animal in the bush!"

He took a step backward and remained tight-lipped with apprehension when Nutsy hopped up my jeans, clung to my shirt, and accepted the peanut I offered her. Nothing that I could say ever altered his opinion, despite the fact that he knew that I was a biologist and that I had spent half a lifetime studying the animals of the North American wilderness.

Time and again I have been forced to listen politely while grossly uninformed people have told me about their imagined narrow escapes in the wilderness. Bears are the number one subjects of these stories, related for two reasons: first because the narrator wishes to seek attention, believing that his humdrum life is somehow made more meaningful by the tale that he is telling; and second because the narrator really does believe that the bear he or she has glimpsed in the wilderness offered a genuine threat.

I remember one man who called me at work one day to tell me that I didn't know what I was writing about after I had done an article on this very subject. He was a telephone linesman, he said, and if he had not been able to escape up a pole, aided by his climbing irons, a female bear would have attacked him when her cub got between him and the sow. I tried to explain to him that a bear could have climbed the pole a good deal faster than he could do so, with or without climbing irons. It turned out that the cub was at least two hundred feet away from where the man was working and that the sow had walked across a clearing to it; keeping well away from the man, the two bears disappeared into the forest. But the linesman was quite convinced that his life had been threatened.

There is no doubt that black bears are powerful enough to be dangerous; if for no other reason, they should be given a wide berth, just as the local farmer's bull should be avoided. Bears have mauled people, sometimes killing them, but investigation of such attacks on adults usually shows that the victims unwittingly have provoked the animals. With children, especially very young ones, it is another matter altogether. Some years ago, in British Columbia, I investigated an extremely tragic attack on a five-year-old boy who, with two companions, was playing on the edge of a forest some distance from the outskirts of the town when a bear emerged and immediately attacked the little boy, knocking him down. His companions ran away, passed right by the local fire station without thinking to ask for help, and waited until they reached their respective homes before telling of the attack. By the time rescuers arrived at the scene, the bear had eaten one of the boy's ears and had mauled him quite badly; the animal was actually eating the thigh muscles of one of the victim's legs when it was shot.

The boy recovered, but he is destined to wear a leg brace for the rest of his life because the quadricep muscles of one of his legs were destroyed.

The bear was young, probably a two-year-old, and was suffering from starvation. This accounted for the attack, yet I am quite convinced that a healthy bear, even one that is not especially hungry, will chase and attack a running child, for it is the instinct of the hunter to run after all those organisms that seek to escape, particularly so when the quarry is smaller than the predator.

Bears, because they are so powerful, must be considered to be potentially dangerous to all humans, a fact that those who travel the wilderness and the parks of the United States and Canada should remember at all times. Indeed, bears encountered in wildlife refuges and other public places are almost certainly more dangerous than their relatives in the wild because they have lost their fear of humans and know from experience that they can often coax food from admiring tourists. Despite the fact that park authorities go to a great deal of trouble to live-trap and remove overly bold bears from public areas, and in the face of the many signs warning people not to feed wild animals or get too close to them, more than a few attacks have been precipitated by incautious people, sometimes with fatal results.

For these reasons, bears should be regarded as dangerous, but I feel that the individual who does not provoke a wild animal and who wisely allows it to go on its way without seeking to approach it too closely has little to fear from either the black bear or any other of the large and powerful wild animals found in North America. Even the lord of them all, the grizzly, who is as powerful as he is unpredictable, rarely attacks without provocation, yet there are many case histories of people who have been mauled and killed by this great beast.

Any animal can be dangerous if provoked; even a chipmunk is capable of delivering a remarkably severe bite, as is a mouse, and yet there are so many people who go out of their way to incite an animal to attack without realizing that they are doing so. Anyone who has visited a zoo and seen some dolt growling at a caged lion or tiger will know what

I mean; in like vein, foolish people make faces and noises at dogs or pick up a kitten by its hind leg. When they are bitten or scratched, there is a great outcry, usually aided and abetted by the media, and all too often the unfortunate and innocent victim of human foolishness is summarily executed.

Although I find myself constantly irked by these things, I can understand them, for those individuals who have spent a lifetime in a city environment and who have gone through a school system that pays scant attention to the living animal but studies its cells or dissects its corpse are genuinely ignorant of the ways and needs of all those vast numbers of living things with whom they share this planet. Ignorance coupled with the human propensity to fear all those things that we cannot understand are largely responsible for the bad press that wild animals have had for so long.

In this respect the flood of publicity attendant with outbreaks of rabies in any given region of our continent makes it look as though the woods are full of mad, ferocious creatures that wait slavering to sink their teeth into the first human being that passes their way. I do not wish to minimize the dangers of rabies; it is a dreadful sickness and one that should be feared by all of us, but it is not nearly as prevalent as the media accounts would have us believe. In twenty-five years I have seen three rabid animals, all of which I have shot: one wolf, one coyote, and one red fox. In that time, I have had intimate acquaintance with countless numbers of others, nursed hundreds back to health, and observed the wild world from Mexico to Alaska; I have been inadvertently scratched and bitten more times than I can readily recall, but never once have I been in danger of contracting the disease. This does not mean to say that one should ignore the possibility of contagion, for it does exist, but to those who would enter the wilderness, I offer the following advice: Learn to identify the symptoms of rabies; avoid contact with any animal that appears to be overly "friendly," respect the rights of those creatures in whose domain you are a guest, and, if possible, learn something about the ways and habits of the animals that live in your own immediate region. Then, *please*, live and let live!

The wilderness is full of fascination, education, and actual value. By observing the natural organisms of our world we may yet manage to avoid polluting ourselves out of existence; we may even manage to learn how to live in peace with our neighbors in other lands.

The more I see of the wild, the more I appreciate its lessons and the greater is my hope for the survival of mankind.

Eleven

A mile or so beyond the northeast boundary of our property, in a region where the forest was largely composed of pines, balsam firs, and spruces, there was a pure stand of full-skirted cedars that arose abruptly and distinctively, rather like an island protrudes from beneath the surface of the sea. This bosk occupied some thirty acres of land; in its center sprawled an untidy jumble of rocks, more than an acre of upheaved granite that disrupted the terrain and climbed to an apex fifteen feet higher than the surrounding understory, the farrago of fractured stone being entirely hidden from view by the encircling conifers.

It was a somber and unusual configuration that testified to the sullen power that lies deep inside the earth, an explosive force that became unleashed during some misty age of prehistory and forced upward a medley of rock forms. Some of these were massive, weighing many tons; others, of lesser proportions, were too heavy to be moved; still more had been sundered into jagged fragments that were scattered far and wide and now lay one on top of the other. To climb over these upheaved monoliths and their attendant slopes of scree was something of an acrobatic feat that was spiced by the risk of slipping into one of the many crevasses that were treacherously covered by mosses and shrubby plants.

Outside of the inhospitable upthrust of granite, hundreds of years of forest detritus had spread a layer of composted soil that attracted and retained moisture, the combination of mulch and wet furnishing ideal growing conditions for the cedars, a species that favors damp places. As a result of these conditions, the unusual enclave of conifers continued to exist in private splendor within a biome dominated by the other species of evergreens.

Although I walked in there often, I usually avoided scrambling over the uninviting rocks, content to work my way through the perpetual gloom of the understory or to sit quietly on some lone boulder so as to watch the animals and birds that found shelter in the penumbral covert. A few mature trees had managed to grow atop the outcrop by snaking their sinuous roots downward through the interstices that separated the larger blocks, in some instances actually embracing the granite and using its weight to obtain better anchorage; many seedlings struggled for survival on the unfertile stone, nourished for a time by the thin coating of soil that had accumulated on level places and in some of the shallow depressions scooped out of the rock by time and weather, but most of these little trees were destined to fall and to die when the weight of their top hampers pulled the roots out of their insecure matrixes; the bodies of many young trees like them, dry and spike-sharp, littered the scree-covered kopje.

When Snuffles was two and a half years old and already largely independent from us, I set out on a morning in late summer intent on visiting the cedars there to observe and, if possible, to count the numbers and species of birds that found shelter and food within the unusual ecosystem. Just before entering the copse I stopped under a large white pine to make a few notes, ending with a reminder that Snuffles had not on this occasion caught up with me, as he so frequently did when his senses alerted him to my presence in his domain. I was fond of the bear and enjoyed his company, but at those times when I wanted to study the wilderness, his presence made my task impossible.

The bear was on the threshold of adulthood and although he still visited us when his sweet tooth reminded him of the good things that Joan's larder had to offer, he now spent most of his time in the forest, a change that occurred quite suddenly in the spring, after he had again spent the winter inside the barn, in his straw house.

Due, I am sure, to the plentiful and well-balanced diet that he had received in his early lifc, Snuffles had become exceptionally large, his girth and stature matching those attained by adult males three or four years his senior. Standing on his hind legs he was a foot taller than I, and when he occasionally became playful and reared up to wrestle, the power of his muscles was alarming.

Believing that I had escaped his notice that morning, I left the shade of the pine and walked over the intervening seventy-five yards to squeeze through the outer screen of cedars, following a game trail that I knew led into the heart of the stand and then split into several lesser tracks. I was not yet halfway to the rock outcrop when the sound of a heavy body moving quickly toward me through the underbrush caused me to stop. Only a bear made that kind of noise, one that was in a hurry and not seeking to conceal its presence.

Moments later Snuffles galloped into view, saw me, and paused so as to make sure that his senses had not deceived him, that the human that was standing there was, indccd, me. Satisfied, he galumphed toward me, weaving dexterously between the tree trunks. He stopped about three feet in front of me, reared up on his hind legs, and peered into my face, his small eyes, framed by a black and hairy visage, glowing like coals, his great paws folded against his chest. I didn't seek to initiate personal contact, but I fished out a handful of peanuts and fed them to him one at a time. As always, I found myself fascinated by the way in which the bear grasped each nut with his lips, rolled it toward his teeth with a big, pink tongue, and daintily bit it open, selecting the nuts but spitting out the shell fragments and the reddish skin. Hitherto, I tried to imitate these actions, discovering that my own mouth was crude and inefficient by comparison. Not only did I tend to crush the entire nut, but its contents, including the fine skin, were immediately soaked by saliva; whereas, when the same task was undertaken by Snuffles or by the raccoons, the shells and skins emerged dry and were easier to discard as a result. When I had given my ursine friend the last nut, I moved on, leaving him standing upright and concentrating on extracting the meat from the husk, an incongruous, gangly figure almost seven feet tall balanced on flat pads.

The bear soon finished his last treat. Still licking his lips, he dropped on all fours and ran to catch up with me, content to amble along at my side as I altered course for the rocks, for now that his company precluded a quiet study of the birdlife of this place, I thought I might as well try to find out what he had been doing when my presence disturbed him. Together we scrambled through the underbrush and young trees that grew on the southern boundary of the rocks, but whereas I was forced to proceed slowly and with caution, Snuffles bounded along easily, his supple body, despite its bulk, squeezing through spaces too low and narrow for me, his seemingly clumsy paws allowing him to dance on the unstable scree with such agility that though he dislodged small avalanches, he was always one jump ahead as the rock fragments cascaded downward. When I had climbed on top of a relatively flat block of granite that was about seven feet above the understory, I paused to look around. The bear, meanwhile, had outdistanced me. From my perch I obtained a good view of those massed rocks in our immediate vicinity, noting numerous disturbed patches of soil, a steaming pile of dung, and several smaller boulders that had been dislodged; all these signs told me that Snuffles had been here for some time and that he was actively exploring the upthrust. I began to move up. As I did so, the bear turned and scrambled higher, but pursued a diagonal course, occasionally stopping long enough to turn so as to make sure that I was following him.

We traveled in this way for perhaps 150 yards, the bear always in the lead and displaying the agility of a mountain goat, I walking and scrambling with care over the treacherous footing, often losing sight of Snuffles because of the shrubs and young trees, or when he dropped into some of the deeper crevasses to squeeze himself through narrow passages before climbing on top of the rocks again.

Reaching an area where the granite blocks were larger and less covered with undergrowth, Snuffles suddenly disappeared between two enormous boulders that leaned toward each other, their upper edges touching and creating a space between them that looked somewhat like an irregular, inverted V. As I scrambled down to investigate I saw that the mosses and fungi that everywhere adhered to the rocks had been frequently disturbed, indicating that the bear was no stranger to this place. At the bottom, which was deeper than I am tall, I heard Snuffles, but could not see him because the entrance to what appeared to be a narrow, upward-sloping cave was pitch black. I became extremely curious, but although I would have loved to go in after the bear, I restrained the urge; I wasn't carrying a flashlight and the place was too dark to risk falling or getting stuck in some crevice. In any event, I wouldn't be able to see, even if I did go inside. Regretfully, I climbed out of the hole and started back to the farm, postponing exploration of the cave for a time when I would come prepared.

Snuffles stayed inside the natural den, his behavior suggesting that he had taken it over for himself as a place where he could rest during those times when he remained inactive, this supposition being confirmed when, on descending the last of the rocks, I found myself close to the place where the bear had met me earlier. He had evidently heard me and had emerged from his shelter to investigate. Would he, I wondered, select the cave as his winter quarters?

When I described the place to Joan later that day, she told me that the week before, walking near the cedars with Tundra, the dog had sought to drag her into the bosk, hackles raised and giving every indication of hostility. My wife, who didn't care for the shadow world within the evergreens and was particularly careful to avoid the granite outcrop, had pulled the dog away, believing at the time that Tundra had scented a bear, as he probably had, this being Snuffles.

From spring until late summer affairs at the farm had progressed peacefully and more or less uneventfully that year. We had accepted responsibility for a cottontail rabbit in May, a tiny, furred creature whose eyes had not yet opened and whom I had named Weavil (because weevils like cotton!), a red fox hit by an automobile that was so badly injured that I had to shoot it, and a few fledgling birds.

We were both grateful for the respite. There were many things that we had neglected to do since the arrival of Snuffles and now that we did not have to worry about him, or about Tundra, who was fully adult and remarkably well disciplined, Joan and I were free to pursue our own affairs.

Manx rarely came near our property now, though I met him on a few occasions while I was walking through the tamarack swamp, an area that the lynx appeared to have appropriated as his range. Slip and Slide continued to live by the lake, but were now less inclined to allow themselves to be seen in daylight; Nose, when last observed in early spring, had moved his range farther north and was, if possible, larger than ever. He recognized me, but kept his distance, and I didn't seek to approach him. Spike and Legs and those other animals who lived on our immediate property continued to thrive, including Babe, Pan, and Pomona. The big doe had delivered only one fawn that year, leading me to believe that she was getting old, for twins are usual after the first pregnancy and until a doe begins to age. Pomona had one fawn, whom we named Diana. Pan, now a splended buck, did not come as often to the maples, but lived in solitary splendor in an area north of our property where there was plenty of shelter and food.

The only cloud that remained on our immediate horizon that summer contained Tundra's hatred for Snuffles and for all other members of his species. The dog now scaled one hundred pounds, a powerful animal who was devoted to us and who was extremely serious about his role as guardian of our property. He wasn't vicious and did not launch himself ravening at strangers, as many dogs will do, but until either Joan or I put his suspicions at rest by saying, "It's O.K., friend," he would prevent visitors from entering the porch by planting himself in the doorway and staring intently into their eyes, his behavior made more sinister by the very fact that he did not bark or growl, or even show his teeth, but conveyed by his entire manner, and especially by the slowly rising hackles on his back, that intruders would do well to wait outside until Joan or I emerged to welcome them.

Nobody sought to walk past Tundra when he stood on guard.

Some of our friends and neighbors were afraid of the dog and would toot their car horns on arrival, remaining in their vehicles until we emerged from the house; others, who knew the dog and were his friends, always were greeted by a great tail wagging and the low, moaning sounds that these dogs make deep in their throats. One neighbor, a local trapper, was always welcomed boisterously because the man kept us supplied with animal meat, saving the carcasses of those creatures that he trapped; Fred was the source of our beaver meat, and whenever he arrived, Tundra announced the fact by venting a series of high-pitched yelps and by lunging at his chain, if he was outside, or by dashing up from the basement when he was out in his pen.

Fred operated a snowmobile in winter, and the previous year, in December, a friend of his borrowed the machine and came with a companion to visit us at Christmas time. Tundra knew both men and allowed them to enter the porch and to knock on our door, but later, when they were ready to go, the dog would not let them touch Fred's machine; he would have pulled them off it if I had not secured him!

Tundra displayed intense animosity toward Snuffles at all times now. When the bear came to visit, we were forced to put the dog in the basement and close the door, at which times he would lunge at the barrier, growl, and yipe his anger. Snuffles, on the other hand, always came in peace and would not be provoked into a fight despite the fact that he weighed an estimated four hundred pounds and could have killed Tundra with a single swipe of one of his massive forepaws. On the few occasions that a confrontation developed because the dog was loose, the bear would turn and run to the nearest tree, or to the house, and scramble out of the malamute's way, there to stay until we had removed the belligerent dog.

For the most part, we controlled Tundra's free travel, knowing his propensity to hunt and not wanting him to revert to a semiwild state, but occasionally he would get away, knowing full well that he was breaking a rule, but unable to resist the temptation. What he did during such unlawful excursions was impossible to determine, but when he came home again anywhere from two hours to a whole day later, he always approached us with ears back, tail down, and dragging himself along almost on his belly, the picture of abject guilt. I would scold him, give him a token scruff shake, then fasten him on his chain, and both of us thereafter would ignore him for a couple of hours, the last being the most distressing to him. He would smile at us, wag his tail, jump, or roll over, resorting to a wide series of tricks in order to try and win himself back into our good graces. Later, when we made up again, Tundra would become delirious with happiness. As well he might! He'd got away and enjoyed himself immensely and now he had the pleasure of returning to the bosom of his pack. Of course, he would run away again given a chance! He knew this, and we knew this; he was not really repentant. But, after all, he was a malamute—and we loved him.

The antagonism that Tundra, like Yukon before him, displayed toward all bears must remain unexplained in the absence of acceptable evidence. I have given the matter considerable thought during the intervening years and I can only guess that it had its roots in ancient, genetic programing stemming from the fact that bears are the only carnivores that can attack and kill wolves, doing so whenever an opportunity presents itself. Although a bear will not seek to attack a pack, knowing that the wolves are too swift for it and that their combined numbers will make it unlikely that the big predator will score, the animal will unhesitatingly invade a wolf den in the absence of the pack, killing and eating the cubs and, if she does not escape, the bitch who has remained behind to care for the young. How such genetic signals are still able to elicit responses from animals that have for centuries been domesticated is a question that I cannot answer. Yet I know that other examples of genetic coding exist and are responsible for many of the behavioral traits observed in animals and even in our own species. Perhaps enlightenment will come one day; in the meantime, one can only guess.

Snuffles may or may not have been typical of all bears, but there is no doubt that he was inherently mild-tempered unless he was hungry and food was denied him. Given those circumstances, he quickly lost his cool, becoming unquestionably dangerous. He had treated us to minor outbursts of hunger-motivated anger when he was still a small cub and had later shown real temper when, on entering the house after his first winter's sleep, he had resisted Joan's efforts to move him from the food cupboard. More recently, we had been forced to forbid him the run of the house because he had become too large and too powerful to restrain, a fact that we discovered the hard way after Snuffles woke up from his second winter's sleep early one evening during mid-March.

It so happened that I saw him emerge from the barn and that, seeking to prevent a confrontation, I removed Tundra from his chain and led him to the basement, telling Joan as I did so that the bear was awake. My wife, not waiting for me to come upstairs again, opened the door and Snuffles entered the house, as before making a beeline for the kitchen and rising on his hind legs in front of the food cupboard. This time, fortunately, Joan didn't try to stop him. She became afraid of him.

Snapping turtle female, the one that injured Sir Francis, the wood duck drake. She was kept for study and later laid three clutches of eggs.

Above: *North Star farm from the air showing, clockwise from lower left: barn and machine shed, storage shed and pump house, gatehouse, and main house. Bottom of photograph faces due north.*

Left: *Joan Lawrence comforting Squirt after the young raccoon had fallen into an empty, forty-five-gallon barrel and almost died of thirst and hunger.*

Joan Lawrence emptying a sap pail into a gathering pail during the syrup season; in the background is the evaporator house.

The bear ripped one of the cupboard doors clean off its hinges and with two more swipes of his paws scattered the provisions all over the floor; he was snuffling around amid this debris when I came upstairs hurriedly in response to my wife's panicky call. When I tried to push him away from a bag of marshmallows that he already had split open, he growled at me, turning his great head toward my legs, his big fangs exposed. It reminded me of the old joke: *Q: Where does an eight-hundred-pound grizzly sleep? A: Wherever the hell it likes!*

Discretion was definitely the better part of valor, yet I could not allow Snuffles to plunder our kitchen. Having prudently backed away from the bear, who instantly forgot his anger while he munched marshmallows, I told Joan to leave the kitchen and to open the front door, after which action she was to keep out of the way. When she had gone, I debated my next move. What would happen, I asked myself, if I let Tundra into the kitchen? My imagination furnished a chaotic answer! No, that was *not* the way to discourage Snuffles; I merely wanted to get the bear out of the house, peaceably; I didn't want to start a war during the course of which our home would be wrecked. At this moment Matilda came to the rescue.

Matilda was a skunk—and I speak in the past tense because I met her seventeen years ago and must conclude that her span of life ended long ago. Yet memory of my experience with her helped me resolve my difficulty with Snuffles!

The skunk, followed by a brood of little striped stinkers, drew herself to my attention in the small hours of a summer morning when I was camping in the backwoods of Ontario. I was asleep, but the sound of something working at the entrance of my tent awakened me. Switching on a flashlight, the head and shoulders of a skunk revealed themselves to my startled gaze, a skunk that was clearly determined to come into my canvas abode. What to do? The first solution that presented itself was to cut my way out of the tent and allow my visitor to investigate the shelter at her leisure. But it was a good tent and I didn't really want to slash its canvas. As I started to ease myself out of the sleeping bag, my right hand encountered a pressurized container of insect repellent; on the spur of the moment, I grabbed the can and squirted a heavy spray into the skunk's face. This startled and incommoded the skunk to such a degree that she backed out of the tent without discharging her own gas; she spent the next half hour sneezing and rolling in the grass outside, her offspring giving her a wide berth.

Sometime later, taking pity on her, I lowered my food pack from where it hung on a tree branch and fed her pieces of bread. She evidently had forgiven me and was a docile visitor once she recovered from the effects of the repellent and, I am sure, from the shock of encountering an organism that possessed a spray at least as effective as her own, if not as smelly. Because she was a nursing mother she was hungry, and inasmuch as I continued to feed her, saving all my leftovers for her and her young ones, she came nightly to the camp during the time that I remained there. I named her Matilda and counted her as a friend when I packed up and departed in my canoe.

Profiting by the experience, I subsequently used pressurized repellent to discourage other persistent and unwelcome visitors, including a young grizzly in British Columbia's Coast Mountains who could not be persuaded to abandon his attempts to get at my tree-suspended food pack. The results of these sprayings was immediate and spectacular and I always remembered Matilda with fondness and gratitude after each experience.

In the face of our bear's greedy and hostile determination,

I gave him a small spraying, an electrifying dose of instant discouragement that had him sneezing and gasping and blundering into things as he sought fresh air outside of our house, there to roll in the grass while kicking with all four legs, bleary-eyed and so disconsolate that Joan took pity on him and came into the kitchen to scold me and to collect the ravaged marshmallows, which she took out to him. But Snuffles had lost all interest in the goodies by then, his sole concern being the removal of the Freon gas that irritated his nasal membranes and caused his eyes to water so badly that he was virtually blind for some minutes. The stuff is almost as potent as Mace, that spray developed to discourage muggers and for the protection of law enforcement officers, letter carriers, and others of similar occupation who must dare the wrath of nasty house dogs. But the repellent is not as long lasting and does not have harmful ingredients. Snuffles was fully recovered about an hour later.

We did not again allow him into the house and all was peace between us from that day on; nevertheless, both Joan and I kept a can of repellent in the porch, just in case it was ever necessary to again discourage the bear or to break up a fight between him and Tundra should such an eventuality develop. Thankfully, our precaution proved needless.

Autumn—and the hunting season—arrived to bless our world with color and to eventually fill it with the sounds of the guns. First it was the time for shooting upland game birds, then the waterfowl became the quarry, and since Sir Francis and Honk had both taken to the airways on their trek south, this period caused us considerable distress.

November ushered in a mild spell and with it the start of the deer season, fourteen days of slaughter that caused me to patrol our boundaries with Tundra as often as I could, a shotgun suspended by its sling from my shoulder, because I had twice been threatened by trespassers who thought that a hunting license also allowed them to aim their weapons at a property owner who challenged their right to hunt on protected land. I was extremely loath to again carry arms in order to protect myself from my fellowman, yet I was not about to allow my property to be overrun by unscrupulous, lawless individuals who would not, seemingly, hesitate to threaten my life. Our property was located a long way from the nearest police headquarters, so I had to be my own law enforcer.

Even so, despite my almost constant vigilance, one hunter got through and gut-shot Babe in the middle of our west clearing. At the sound of the rifle, I jumped on the tractor, forgetting to carry my own gun, and raced away, arriving at the scene just as the man put a last bullet into Babe's head. My anger was fanned by horror; I was glad later that I had left the gun at home; I never carried it again.

The upshot of this illegal and unprincipled action was that I made a citizen's arrest on the spot, telling the man that he had two choices: he could attempt to evade justice by escaping, in which case he would add a fugitive count to charges of trespass and the discharge of a firearm on posted land, or he could consider himself under arrest and wait until the police arrived to formalize the proceedings. He chose the second alternative and was duly arraigned, found guilty, and fined one thousand dollars. But the irony of the case was that the law, while meting punishment for the violations and taking into account two previous convictions for the same offense, allowed the poacher to keep the dead deer on the premise that he had legally obtained a license to shoot the animal.

Why, I often wonder, do so many men feel that they must go out once a year and destroy some lovely, living thing? Where is the pleasure, or the sense of prowess, in watching a pair of keen, velvet eyes become dull and unseeing, while blood spouts out of a horrible wound, and the rictus of death moves the bowels and the bladder? In the face of need, I can justify hunting; I have done so myself. One must eat. But in a society where food is abundant and so readily available, hunting for so-called sport is a barbaric and murderous activity. It used to be that a few unprincipled men earned for all hunters a reputation that most of them did not deserve. In more recent times the opposite is taking place; there are so many ignorant, brutal, lawless men stalking the forests in the hunting season that the good, careful hunters, those men who take pride in the stalk and in effecting a clean and instant kill, are so greatly outnumbered that many of the ones that I know personally have put their guns away, disgusted with the annual carnage and feeling that it is no longer safe to enter the wilderness at this time of madness.

That was a sad autumn for Joan and me. We avoided the west clearing for weeks, until the snow had fallen to decently cover the spot where Babe had died; even then, memory of her shocking end remained alive.

Pan survived the carnage; so did Pom and her fawns, and Diana, but we were glad when the five deer wandered away into the deep wilderness for the winter, despite the fact that a wolf pack moved into our area in December.

Snuffles disappeared some days before the waterfowl season opened, but I had been too busy to visit the cedar woods to see if he had bedded down in the rock cave, which I went back to explore a week after Snuffles first led me to it. The den was composed of a chamber about five feet high, six feet long by just under five feet wide, a roughly oval grotto reached by a short, upward-sloping passageway. This tunnel, starting at the inverted V entrance, where it was high and wide enough to allow me to enter by stooping only slightly, narrowed dramatically as it advanced into the granite, its terminus allowing only just enough room for me, or for a bear, to crawl into the end chamber. It was an ideal den and showed many signs of continued occupancy. Insofar as it was only about a mile away from where I found Snuffles crouching beside his dead mother, I wondered if this had been his birthplace. At the far end of the cave was a mass of old leaves and grasses, scattered now, but clearly the bed of a sow—males, or boars as they are also called, do not often bother to scrape up forest duff so as to make a bed. By the same token, the den might well be considered the private winter domain of some other bear, in which case Snuffles would be chased out if its rightful owner was large enough to evict the trespasser.

The snow was quite deep on the ground when Joan and I left the house with Tundra to go and see if Snuffles was, indeed, asleep in the rock cave, though my wife had made clear before we left that she would remain with the dog outside the cedars while I went in to investigate. This was a sensible precaution, for I wasn't anxious to lead our dog to his sleeping enemy, but my wife was motivated not so much by reason of the dog's presence as by virtue of the fact that she always felt uncomfortable inside the gloomy understory; she didn't know why this was so and she didn't question it; neither did I, though I sometimes wondered if she suffered from slight claustrophobia.

It was a nice, sunny day with the barometer registering two degrees below zero as we set out shortly after breakfast. I held the dog's lead, leaving Joan free to feed the chickadees that accompanied us as we traveled, a happy flock that never failed to materialize from autumn to spring every time that either one of us stepped out for a stroll, but that ended almost abruptly with the arrival of the mating season, and with the coming of the insects, the staple, warm-weather food of the black-capped little birds.

A couple of hundred yards before we reached the cedars I handed Tundra's lead to Joan and suggested that she keep the dog at least that distance away from the evergreens to preclude the possibility of his getting the bear's scent. Even so, Tundra's behavior suggested that, if not a bear, some exciting effluvium had reached his keen and twitching nose and he did not take kindly to being left with Joan as I headed for the rock outcrop, flashlight in hand.

A quarter of an hour later I stood before the doorway of the den, bent low and sniffing the rancid smell of recumbent ursine. This told me that the cave was definitely occupied; but was it Snuffles in there, or some nasty-tempered old gentleman who might rush me as I was halfway along the narrow tunnel? Hesitating, I could not help recalling the last time that I had done something like this; the memory of that adventure in the mountains of British Columbia did not help me now to find the courage I needed.

A friend, who was a conservation officer in charge of a wilderness region in northern British Columbia, invited me one day to inspect a denned grizzly, which was something that I had repeatedly told him I would like to do, especially if I could get a photograph of the sleeping giant. He had promised to let me know the next time that he found a den; one January day he telephoned me and offered to conduct me to the place. I was elated! Together with camera and electronic flash unit I presented myself at his house and we set out in a government four-wheel-drive vehicle, climbing steadily for several miles over a rutted logging road until we reached a valley area where the track ended. Traversing the half-mile of flatland on snowshoes, we came to a section of mountain that sloped gently and was littered with dead trees, this being a talus section on which the trees could not cling successfully. At a place where a number of downed trees had mounded against a large rock, a cave had been formed and as more breakaway rocks and trees accumulated, it had developed into a cavity large enough to shelter even the largest bear. It was approached by a short, narrow, low tunnel the mouth of which was not much more than thirty inches in diameter. The entry was covered in snow, but a vent hole was noticeable, its presence due to the exhaled breath of the occupant, which was just warm enough to prevent the snow from totally concealing the den mouth. As is usual, the stale breath had turned the snow quite yellow immediately around the vent.

Helped by my friend, I excavated the entrance and was heartened to note that it was so small, believing that only a young grizzly would be able to squeeze through it. In due course, camera around my neck and flash unit in one hand, I crawled in, moving very slowly and taking care to make as little noise as possible, my progress aided by my friend, who was bent low and holding a flashlight. All too quickly I came to the denning chamber, my nostrils filled with the acrid smell of bear; my friend's flashlight could not now reach inside the grotto to illuminate the occupant, so I fumbled for a small, two-cell light I had in my parka pocket, got it out, and risked projecting its beam in the darkness ahead. Sight of the largest grizzly I have ever seen caused me to quickly extinguish the light and to remain unmoving, my heart pounding so violently that it seemed to be climbing into my throat. The bear continued to snore, and after a couple of minutes I took courage and switched the light on again, aiming it so that the bear would not be in the center of the cone. The animal was curled up, nose to tail, its flanks heaving with each slow breath that it took, the exhaled gases turning to vapor within the cooler atmosphere and becoming visible in the light. He truly was a large bear (I *think* it was a male), an intimidating creature despite his air of repose.

I began to wonder what would happen if he should suddenly wake up and rush me. Then I fervently regretted thinking about such a possibility. To put an end to my hesitancy, I forced myself to think about the photograph that I had come here to take and because I could not manage to focus the camera one-handed, I placed the flashlight on the rough floor of the den, aiming it at the grizzly's rump by propping the lens on a small, flat rock. I began to focus, got the creature in clear perspective, was about to push the button that would simultaneously activate the flash—when I chickened out! It suddenly occurred to me to consider the effects of a high-powered flash inside that confined chamber: if that sudden burst of brilliance didn't wake up the grizzly, nothing ever would!

Cravenly, I turned myself around and retreated, my back crawling and my adrenaline flowing as I inched my way toward that blessed daylight up ahead, expecting to hear at my back at any moment the rush of one large and angry grizzly.

Outside, my friend wanted to know why I hadn't taken the picture with the flash. I told him I hadn't taken a picture, *period*! And I told him *why*. He grinned superiorly, taunted me. I was a chicken, a supposed outdoorsman who was afraid of a sleepy old grizzly. I kept nodding yes. It was all definitely true. When he finished, and while he was still smiling broadly, I removed the camera from around my neck and extended it to him.

"Here, you show me how it's done!"

That simple sentence wiped the superior smile right off his face; it even caused him to back away from the den mouth. Perhaps more important as far as I was concerned, it effectively prevented him from kidding me in public about the affair, though he had many things to say about my fear when I was not present to reciprocate in kind. I have said, and I still believe, that adventure is the sum total of something nasty happening to somebody else. But that encounter with the grizzly was an exception. That, *most definitely*, was an adventure happening to *me*, personally!

Standing before what I hoped was Snuffles's den that morning, I didn't do my morale any good by recollecting the grizzly in British Columbia, but on the premise that there was nowhere to go but in, I stooped, passed through the portals, and soon was crawling on hands and knees, the flashlight held steady in my mouth. Two or three minutes later my head penetrated the chamber and the light revealed the recumbent shape of Snuffles, who was lying on his chest and stomach, big head tucked between extended front paws, his haunches turned so that his back legs stuck out to the side; he was fast asleep, and snuffling.

Winter was severe that year. It came soon after the end of the hunting season and brought with it several blizzards and a cold that caused the mercury to drop to thirty-five degrees below zero. By the third week of January there was almost four feet of white on the ground, but the blizzards had left the land enveloped in quiet, bathed by the cold sun during the short days attendant with that time of year.

In the wilderness the animals and birds sought shelter in the deep forests, hunter and hunted alike, and even those of our wild friends who lived near our buildings came less often to visit. Penny, of course, had denned in the late autumn, as had all the raccoons, but Spike and Legs remained with us, and Weavil, now having attained full size. Twice during January Boo came in daylight and was glad to receive a free meal on each occasion. Our resident birds, the blue jays, nuthatches, woodpeckers, and chickadees came more frequently to the feeders and fat stations, a fact that caused us to use nine hundred pounds of seed before spring arrived to make the wilderness warm again.

Twice I loaded bales of hay on the wagon and hauled this by tractor the tortuous mile and a half that separated us from the place where Pan and his sister and other deer were sheltering. I didn't see them on either occasion, but the hay was eaten. On my second journey I was concerned to find wolf tracks in the area and when I followed the spoor on snowshoes for about three miles I found at trail's end the remains of one of the fawns, something I did not have the courage to tell Joan when I returned home.

The pack consisted of seven wolves. I had seen them on a number of occasions as they loped through the wilderness and we heard them howling often, their mournful calls reaching Tundra and causing him to become extremely excited, though he did not reply. In this respect he was quite unusual, the first sled dog I had known that did not howl often, especially when the wolves called. Howling is a characteristic of northern dogs, but Tundra did not share it.

Preceding one of the wolf sightings, I was fortunate to hear them barking shrilly while I was still about half a mile from where they were, the stationary quality of their latration suggesting that they were engaged in some exciting pursuit that kept them bunched in one specific area. It was snowing lightly, the kind of brittle, frozen flakes that feel like grains of sugar and can sting mildly when the wind causes them to pelt into the face. Because the gusts were coming from the direction of the wolves, I quickened my pace, hoping that I would get the chance to approach them unobserved and thus be able to study them.

As I snowshoed, I guessed that the pack was on one of the beaver ponds, but I could not imagine what they were doing that would cause them to remain there continuously and to bark with such abandon and with such evident excitement. Wolves bark readily enough, but these outbursts usually are not sustained for long unless the animals are alarmed, or made frantic by the close proximity of a prey animal. In such cases the calls are different, tending to be uttered continuously and being of deeper tone; what I was listening to as I hurried toward the frozen pond had a different timbre, a higher pitch, intermittent and staccato latration reminiscent of the sounds made by a group of juvenile wolves at play.

Approaching the environs of the pond I realized that the wolves were calling on the move, the sound of their voices going away from me at one moment, coming toward me the next. To make my final approach I removed the snowshoes and went forward with extreme caution, maneuvering myself so that I eventually obtained a good view of the lake while being concealed by a group of balsam firs that grew atop a slight knoll.

The pack was running in loose formation 150 yards out on the ice. The wolves were playing an intense game, the object of which seemed to be to catch up with one exceptionally large individual, the leader, I guessed, who dodged and leaped and turned and often charged individuals, knocking them off their feet kicking and struggling. They were all having a great time, like puppies at a kennel allowed loose for an outing.

Through the glasses I noted that four of the wolves were adult and three were juveniles, gangly ten-month-olds who were making most of the racket. The big wolf was black and carried a distinctive white blaze on his chest, an animal larger than Tundra, more akin to Yukon in size, and rather unusual in a region where wolves are predominantly smaller than their northern brethren. Then and there I named the leader Lobo, which means "wolf" in Spanish and thus is not very imaginative, but it would do for my purposes.

Watching those seven magnificent wild dogs, it was impossible to fault them for killing the fawn, sad though I was over her death. For half an hour the pack played on the frozen pond, then Lobo stopped practically in mid-stride, stared directly at my place of concealment, and turned swiftly to lope across the lake, leading his pack and soon disappearing into the forest. For a second or two I had looked directly into his eyes, the field glasses bridging the distance between us, and although I could not determine what it was that had warned Lobo of my presence, I knew from his look that I had unwittingly made some move, or sound, that had reached his ever-keen senses.

Following that accidental encounter I spent a great deal of time studying Lobo's Pack, as it was designated in my notebook, tracking the wolves, listening to them howl, following them for many miles in the deep wilderness, and actually sighting them on a great number of occasions. The wolves knew I was traveling in their domain and, as I had experienced before with other packs, they lost their fear of me when I offered no threat. They remained cautious, of course, but toward the end of that winter I was able to approach them quite closely, sometimes seeing all of them as they trotted casually into the shelter of the forest, at other times sighting one or two of the pack members. The duration of these sightings varied. More often than not I was allowed only brief glimpses as the sinuous animals faded away, but there were more than a few occasions when I was allowed to watch them for several minutes at a time, especially Lobo. The big leader repeatedly would leave the pack to circle, get behind me, and follow at a distance of fifty or seventy-five yards, usually remaining unseen, yet allowing me to catch sight of him often enough. He was clearly as interested in me as I was in him and his pack.

Toward the end of February I took to carrying a beef knuckle bone or two when going wolf watching and the first time that I detected Lobo's presence after that I dropped the offering on the snow and continued to follow the tracks of his group. On my return via the same route the bone was gone and the wolf's big tracks as well as the impression of his chin and nose in the snow, one indentation on either side of the place where the bone had landed, told me that he had accepted the gift. This now became a ritual; later it led to a closer acquaintance with the big wolf.

The winter was all too short for me that year, and with the coming of spring and the melting of the snow my wolf study was interrupted; not only was it next to impossible to track the pack over the bare ground, but I knew that at least one of the bitches, the dominant female, would soon be whelping, as a result of which happy event the pack would seek a quiet place, far from the haunts of man.

As always, we were glad to see the advent of spring and to meet up with those of our wild friends that had slept through the rigors of winter. And, of course, it was birth time, the glad season of new life that rewarded us for our care of the wildlings we had befriended by allowing us to see the young animals that would not have been born without the assistance that we rendered their parents.

The maples once again responded to the warmth and offered their sap for the taking. We became busy, the days passed. One evening we were finished and had time to wonder about Snuffles, who had not yet made his presence known. We talked about his absence, half-hoping that he wouldn't come back because he was now beginning the fourth year of his life and needed to become fully independent, yet also wanting to see him again, a desire that would not have been approved of by Tundra had he been aware of it.

Twelve

When April ended without revealing any signs of Snuffles, we concluded at first that he had returned to the wild following his third winter's sleep, and we were somewhat thankful; but after we spent several days during early May looking for him in the forests, our relief turned to worry. During the last search I took Tundra with me, thinking that if the bear was in our neighborhood the dog would not fail to locate him, but after spending an entire day scouting some fifteen square miles to the north of our property without finding so much as one fresh mound of his dung, I was forced to believe that Snuffles had either strayed far away from our area or had met with an accident.

We had faced this sort of dilemma quite often during the years that we had devoted to the care of wild animals, but we never became used to its uncertainties. Despite the fact that we really did want our wards to gain full independence, we worried about them until we had proof that they were sucessfully established on suitable range; in the absence of this, we fretted, sometimes for months.

Wild animals are capricious and tend to follow their noses as they wander in search of food and shelter, the duration of their peregrinations depending largely on the needs of each species as well as upon the amount of competition exercised by their own kind. As a general rule, small animals make do with a relatively few acres of land, but the large ones, especially the predators, need many miles of range on which to hunt. The snowshoe hare, for instance, lives contentedly (in the absence of overpopulation) on about one hundred acres of forested habitat, while the timber wolf requires a territory with a radius of at least one hundred miles. Bears, although much larger than wolves, occupy less range because they are omnivorous, the females foraging over an area that may encompass ten miles, and the solitary males, unhampered by young, occupying about twenty miles of range.

In most cases we eventually would discover the whereabouts of our former wards, but some of the animals that we raised disappeared after their release, never to be seen again and leaving us to worry about them for an indefinite time. Birds, particularly the migratory species, rarely remained in our region once they were able to fly; they always proved to be far more independent than the mammals and less prone to remember their association with us, characteristics that were not surprising in view of the many differences that exist among the two classes. It was indeed a *rara avis* that continued to favor us with its friendship; that was why we tended to spoil Maggot and Boo whenever they came calling.

Spring was kind to us that year. It came peaceably, ushering several weeks of exceptionally fine and warm weather that allowed us to perform a variety of tasks neglected during autumn and winter. Afterward, refusing to become overly depressed by the absence of Snuffles, Joan and I devoted a good deal of our time to Tundra, who was in seventh heaven now that he did not have to compete with his enemy.

Before we began to search for the bear, the dog and I undertook a number of long journeys into the hinterlands, my object being to train Tundra to stay at my side while not secured by his lead, as well as to convince him that carrying a backpack was a duty that most northern dogs were only too glad to assume. I enjoyed moderate success in the first intent, but failed dismally in the second. Our malamute was not at all disposed to shoulder a burden! Every time I fastened a pack to his protesting body he became rebellious, tugging at his lead, shaking, and jumping up and down, sometimes tangling himself between my legs. Half a mile of such going was as much as I could stand without becoming enraged and being seized by the urge to take a club to my companion. I didn't, of course; that is *not* the way to get the best out of any living thing, but each experience was so frustrating that I eventually admitted defeat. The dog was happy to pull a sled, but packing was just not his thing; I had to accept that.

Hunting was another matter! Like Yukon before him, Tundra could not resist the abandoned joy of the chase, his fine senses leading him unerringly to all manner of animals and birds. This forced me to watch him intently when he was free to run, but, drawing on past experience with Yukon, and prepared by a canine "body-language" chart I had once compiled to help me interpret the wolf-dog's reactions to environmental signals, I was able to turn the malamute's predatory urges to advantage, putting him on his lead whenever he showed by his behavior that something of considerable interest had alerted him, then allowing him to guide me to the quarry. In this way, controlling his charges and advances as quietly as possible, Tundra discovered for me many organisms that I otherwise would not have detected. Following a dozen or more of these encounters, although the dog would have preferred to chase and to kill the quarry, he began to enjoy the game and learned to work with me. Now, instead of tugging at his lead, he would pause, look at me, and set off at a stalking pace, preceding me patiently and often waiting so that I might negotiate some obstacle that to him presented no difficulties. We became a team engaged in a bloodless sport that offered him the thrill of a moderate "hunt" and allowed me to begin a count of the many life forms that inhabited our area of wilderness.

I began to map our land and several miles of wilderness abutting it, compiling an ecological blueprint that showed ground contours, composition of soil and rocks, vegetation and water sources; later, using an acetate overlay, I plotted the territories of all those species that I had encountered, marking dens, nests, and bedding places, an ongoing study that necessitated frequent changes and additions, but which soon began to reveal some fascinating and valuable information. Engrossed with my work, I stayed up late one night to complete a new plot and to prepare an overlap chart showing an area four miles north of our land. Joan was sound asleep. Tundra lay under my desk, awake, but fully relaxed; outside, a full moon bathed the wilderness with inviting light. My watch read 2:15; the silver night was too alluring to resist. I was about to rise, intending to take Tundra for a walk before seeking sleep, when the dog jumped to his feet and dashed into the kitchen, hackles up and lips peeled back to reveal his formidable fangs. When I caught up with him he was standing on his hind legs, forepaws against the edge of the counter, glaring into the visage of Snuffles. The bear was also standing upright, his own front paws resting on our window feeder. He was looking into the house as intently as Tundra was looking out of it. I went upstairs to wake Joan and to tell her the news.

Our reunion with the bear, whom we did not allow into the house on this occasion, was not as boisterous as last year's, but it made up with dramatics what it lacked in rowdiness. Snuffles showed himself less ready to become intimate and although he eagerly accepted a honey sandwich that Joan offered him, he growled at her as he swallowed the last mouthful when she extended her hand to pat his head. This unexpected rebuff caused her to back away from him quickly, her eyes round with apprehension, quite forgetting that she was holding an open package of marshmallows. Snuffles reared onto his hind legs, towering over my wife, then shuffled toward her menacingly, his small eyes fixed on the sweetmeats and his nostrils twitching as he snuffled. Joan retreated some more.

I stepped between them, took the package out of my wife's hands, and scattered its contents on the ground. Snuffles dropped onto all fours and began eating his favorite treats. That night I told Joan not to feed him again.

The bear was more than three years old. He was too large and too strong to be allowed to hang about the vicinity of the house, not only because of his relationship with Tundra but also because he could become dangerous if denied food. Discussing this later that night, in our bedroom, Joan confessed that she had experienced intense fear when Snuffles growled at her, and although this emotion was readily understandable under the circumstances, her dread upset me considerably. Animals are quick to detect fear. Given the fact that predators are prone to attack those organisms that seek to escape from them, Joan's nervousness, were it to continue, might one day precipitate a tragedy of unthinkable consequences.

We awakened late next morning. The day was warm and sunny, a glad morning, yet I could not shed the feeling of apprehension that had seized me when Joan told me of her fear of the bear. I had, of course, noticed her initial apprehension when she retreated from him, but I read it incorrectly, believing that it was a passing moment of alarm. But it had been more than that; the bear had seriously intimidated her, and she had allowed him to detect his dominance, circumstances that could well lead to another confrontation. This was worrying, but I took care not to show my concern in case it intensified her nervousness. It would be best for all concerned, I reasoned, if Snuffles could be coaxed to go away again, thus allowing my wife the time to get over her anxiety.

I went outside to look for the bear. He wasn't to be seen near our premises. Wondering if he had elected to return to the wilderness of his own accord, I thought I might as well go and look for him, taking Tundra along to facilitate what could otherwise become a daylong task.

Putting the dog on his lead, I instructed him to "find Snuffles," a command that he was quick to obey because he readily recognized the name of his foremost rival. He needed no urging to start looking, casting about eagerly, leading me twice around the house and halfway to the maples before he found the freshest scent. Now, nose to the ground and tail curled up tightly, he set off purposefully, aiming toward the northeast. An hour later we had covered about three miles, following a meandering trail through heavily wooded country during the course of which I had seen a number of bear tracks and other signs that suggested that Snuffles was feeding as he traveled. The bear clearly was working his way into the deep wilderness; this was encouraging, but I decided to follow for a time yet, to make sure that he really was locating himself some distance from the farm. We spent one more hour dogging the spoor, after which, estimating that we had covered about six miles, I turned around and retraced our route, satisfied that at least for the time being Snuffles would entertain himself in his proper environment.

The next few weeks passed uneventfully. Snuffles didn't return, and Joan appeared to have forgotten all about her fright as she once again applied herself to the business of gardening, this time selecting a patch of ground nearer to the house, where Tundra could keep an eye on the vegetables and so discourage the raiding groundhogs. By mid-June, a month after the bear had come and gone, I became nearly convinced that he had at last become independent.

It may be that I overreacted to the incident between Joan and Snuffles, but many years of wandering through the wilderness had taught me that one should never take chances if safe alternatives are available. It is one thing to risk injury, or even death, if one must do so in order to survive, but to be forewarned of danger and to continue toward it on a possible collision course is the pursuit of fools; one *may* get away with it, but it has been my experience that luck frowns on those who do not weigh the odds before they act.

Knowledge of wild animals in general and of Snuffles in particular suggested that our bear would not deliberately hurt Joan under normal conditions, but the fact remained that he was powerful enough to injure her severely, even kill her, if he was provoked. On all fours he was more than six feet long; standing upright he passed the seven-foot mark. He probably weighed 450 pounds by then, and his armament was quite formidable: tusks that were more than an inch long, heavy, powerful claws that measured two and a half inches in length, and massive limbs. Inherently gentle, he could be boldly aggressive when his mind was fixed on food, and after seeing him turn over with one paw rocks that were too heavy for me to even *move* with both hands, I had no doubt about his strength. Nevertheless, I was nearly sure that in the absence of fear, and provided that we did not tempt him to become aggressive by allowing him to detect food that we were not prepared to give him, Snuffles would not seek to injure either one of us.

Fear and aggression produce the same physical manifestations, the essential difference between the two emotions stemming from the way in which an animal interprets environmental stress. What may elicit aggressive behavior from one individual may well cause another of the same species to become afraid; in either event, the neural and physical responses are identical in each case because the bodies of both organisms must instantly be made ready to deal with the emergency. The complex series of reactions that are triggered when an animal is alerted by abnormal changes in its environment are collectively referred to by biologists as the *alarm response*, but I believe that a better term would be *response to challenge* inasmuch as it makes no difference whether an organism is motivated by alarm or whether it has been urged to attack. The main thing is that it is reacting to a challenge.

Aggressive and defensive behavior in all vertebrates begins when the senses detect some unusual situation. With unmeasurable speed, information is telegraphed to the brain by means of nerve impulses. If an emergency exists (that is to say, if attack or flight are indicated), the body must be instantly prepared to handle sudden and intense physical activity. All these things occur in sequence, but so astonishingly quickly that it appears as if all reactions were taking place simultaneously. The end result is that a series of glands referred to collectively as the endocrine system are commanded to step up their output of specialized hormones ("chemicals") that are always present in the blood in regulated amounts. What follows can be described as an instant "shot in the arm," principally triggered by the adrenal glands, one part of which, the *medulla*, injects directly into the bloodstream a surge of a hormone called *epinephrine* (also known as adrenaline). This substance immediately increases the heartbeat, steps up breathing and blood-clotting rates, causes blood pressure and blood-sugar levels to rise, increases muscle tension, and produces resistance to fatigue. All these reactions are essential when peak performance is demanded from any vertebrate system, including that of man.

Thus, if a deer becomes aware that a wolf has selected it as prey, it will react defensively by seeking to escape. (It has become afraid.) The wolf, sensing the deer, behaves aggressively. (It is urged to attack.) Each animal is motivated by a different emotion, but their bodies are responding to nearly identical mechanisms; they are, in effect, powered by the same fuel.

Although predators are programmed to attack, they don't always do so; even when they are hungry they first assess their chances of success and balance these against the risk of personal injury. If their keen senses notice that the prey animal is likely to put up a stiff fight, the experienced hunter will back off, as wolves do when a moose turns at bay; but if the intended quarry shows by its behavior that it is afraid, an attack is almost certain. In the absence of either fear or aggression, even the largest and most powerful predator will become unsure of itself. The would-be quarry, by exhibiting neutrality, is creating doubt. Its behavior cannot be predicted: it may run if pressed, or it may fight if attacked; it may even become the aggressor. Such uncertainties undermine a predator's confidence and usually prevent an attack, for any animal that depends on its own physical prowess in order to keep itself fed cannot afford an injury. In order to survive, it must always seek to keep itself in peak condition; even a minor wound can reduce efficiency, while a serious one will almost certainly cause it to die of starvation, or else leave it too weak to defend itself against another predator. There are exceptions to this "rule of thumb," of course, just as there are differences in individuals, but I believe that it generally holds true.

My experience with wild animals has taught me that to be neutral is to be safe and, conversely, that to be afraid is to invite attack. Hence my anxiety when Joan exhibited fear and Snuffles became bold enough to advance on her, seeking to take by force the marshmallows that she was holding. Today, with the advantage of hindsight, I recognize that we should not have encouraged the bear to hibernate in the barn during the second winter of his life. He was then old enough and large enough to survive in his proper habitat and would have found for himself a suitable den—as he did the next year—in which to spend the cold months. By allowing him to stay with us and, later, by making a fuss of him and feeding him luxuries, we invited him to delay his independence without considering that he would one day become too big and powerful to be trusted. Had we raised him to be a pet, the risk factor could have been reduced to a certain extent, but only for a relatively limited period, for, like all essentially solitary animals, bears do not adapt well to social conditions and tend to become short-tempered as they grow older. Sooner or later such animals must be disposed of, either by placing them in a zoo or by destroying them before they are driven, through no fault of their own, to commit an unforgivable trespass. Certainly, were I to become foster parent to another bear, I would not repeat my mistakes.

The most charming and docile of all the wild animals that lived in our neighborhood was a four-ounce little acrobat who had large red eyes, fur like fine velvet, and "wings" that allowed him to glide from one tree to another. He was a flying squirrel, a member of an almost totally nocturnal species that discovered our window feeder one night and returned thereafter with several companions.

We called him George. He was the friendliest of them all and he showed himself to be quick to grasp the relationship between us and the seeds that he came to eat. We noticed him for the first time at ten o'clock during a night in winter when his eyes captured the light from a table lamp and returned deep ruby reflections. Silent as a ghost and quick as mercury, he scurried nervously over the feeding platform, his body a pastel blur, stopping as quickly to nibble at seeds for some moments, then to scuttle away again, sometimes launching himself into the darkness and remaining absent for several minutes at a time, but reappearing as a new blur that landed soundlessly on the tray.

We watched for him the next night and he didn't disappoint us, arriving within minutes of yesterday's time, but now accompanied by four of his relatives. Unlike the red and gray squirrels, who are solitary and uncompanionable except at mating time, the flying squirrels *(Glaucomys volans)* are a sociable breed, actually enjoying each other's company and congregating in complete harmony. Although of nervous disposition, they are remarkably tame if they are unmolested and given sufficient time to acclimatize to the ways of humans.

Knowing these things, we did not seek to intrude on the inquisitive little elfins when they came to feed, being content to watch them from a respectful distance while they spent at least half an hour dining on the fresh seeds that we made sure were put out for them every night. They were remarkably punctual, coming within a few minutes either way of 10:00, staying until 10:30 or 10:45, and leaving one by one as silently as they arrived. Years earlier I had been greatly intrigued by the findings of one biologist, Patricia DeCoursey, who selected the flying squirrel as the subject for her doctoral thesis at the University of Wisconsin. During a series of remarkable experiments over a period of several years, DeCoursey established that this small rodent has an extraordinary sense of time by means of which it regulates its periods of activity with the onset of darkness, returning to its den before the coming of day. DeCoursey's studies were conducted mostly in the laboratory and contributed important information relative to the biological rhythms that govern the timed movements of most, if not all, terrestrial organisms; understandably, the researcher was not able to follow flying squirrels during their active time in the forest, so her findings did not seek to establish patterns for animals in the wild. With these things in mind, I began to study George and his companions, sometimes setting the alarm to wake me every two hours, at other times maintaining an all-night vigil beside the kitchen counter.

My findings were interesting, though not conclusive, owing to the irregularity of my vigils. Yet I was able to establish that George and his friends came to feed four times each night, their first visit made at approximately 10 P.M., their second at 12:30, the third at 3 A.M. and the last at 5 A.M.* What they did between visits I cannot be sure of, but after a number of accidental encounters in the forest night, and, in winter, aided by the marks of their landings in the snow, I am inclined to believe that they continued to forage in the trees and on the ground when not occupied at

*These times are not precise, but varied between three and twelve minutes either side of those indicated, the mean average for the winter being three minutes forty-five seconds.

the feeder. On those occasions when I cleaned all but a few seeds from the feeder in between their visits, the squirrels came at more frequent intervals, until I again offered them a good supply. Like Joan's bees, George and his relatives proved that they did not need mechanical clocks to help them keep an appointment.

By spring the delicate, fawn-colored night squirrels trusted us sufficiently to accept peanuts from our fingers, and George would slide down a tree trunk in order to lap peanut butter from the end of my index finger, always provided I was alone and that Tundra was either in his pen or chained outside the porch. I discontinued timing the squirrel and his companions when the maple sap began to flow; afterward, the arrival of Snuffles, and Joan's reaction when he growled at her, caused me to delay the study still further.

Some days after the bear had returned to the forest I left the house alone, intent on examining the cedar stand and the rock upthrust, curious to see if Snuffles had visited the place since awakening from his winter's sleep. On the way there, in the smallest and most northerly of our clearings, I put up a sow bear and her two cubs, twin replicas of our own bruin at that age.

The sow was grazing when I disturbed her, cropping the mixture of clover and alfalfa that had been seeded there years earlier by the previous owner but which was now rather sparse and straggly. The cubs were mooching around near their mother, playing, it seemed, and a little slow to respond when she grunted her warning. She was a big bear, an elderly animal still in her prime: and when she rose on her hind legs to look my way, she appeared to be even taller than Snuffles.

I had stopped on sighting her, and because I did so she wasn't in a great hurry to leave, but she became impatient with the cubs when they failed to obey her command. Dropping onto all fours, she rushed one of the youngsters and boxed his ears, sending him rolling over and over like an animated black ball. The cub squealed loudly, hastily found its feet, and ran for the trees, closely followed by its sibling, who was clearly anxious to avoid getting his share of the punishment. The sow went after them at a stately trot, and the three soon disappeared from view.

This encounter explained the absence of our bear. Sows with young will not tolerate a boar in their territory—and with good reason, for if a cub has a major enemy, it is an adult male. Fatherhood is a responsibility unknown to the ursine species. Solitary by nature, dominated by a large appetite, and devoid of paternal instincts, boars roam the wilderness continually searching for food and will unhesitatingly kill a cub if its fiercely protective mother is not there to defend it, a characteristic that is difficult to account for accurately. Most mammalogists suppose that the bear's cannibalistic appetite has been programed into the species by natural selection as part of a safeguard against overpopulation, a plausible theory when it is considered that—in the absence of man armed with modern weapons—disease or another bear are the only two dangers that beset the young of the species. There are other examples of this method of population control to be found in nature, not the least of which is seen in the shark, the females of which species devour their own young. The foot-long babies, hatched within the mother in most breeds, and fully equipped to survive as soon as they are born, waste little time in seeking shelter as soon as they emerge—I have actually seen a female shark producing young and at the same time snapping up her own offspring when these ventured too close to her jaws. Undoubtedly a lot of young sharks are killed in this way, but those that survive are tough and resourceful creatures that usually make it into adulthood, from then on being almost impervious to natural predation.

Under the circumstances, I was glad to see the sow and her two cubs, her presence just about guaranteeing that Snuffles would keep away from our neighborhood while the bear nursed her young ones, confining them to a relatively small range until they were older and better able to travel.

I arrived at the rock upthrust to find that the old lady bear had usurped the territory, perhaps using it since winter as a suitable bedding place for herself and her cubs, whose small tracks were evident on the forest floor and among the rocks, on those flat places where the thin soil had been disturbed, probably by the playful cubs. As I extended my search of the secluded area, I found further evidence to suggest that it had been in constant use for several weeks: things like droppings—those of the cubs easily distinguishable from those left by the sow—branches chewed and clawed by the young bears, and rocks recently turned over by the mother as she searched for insects.

It was late afternoon by the time I finished my examination of the cedar world and started back for the farm along a trail that would take me to the east of the clearing where I had surprised the sow. With any luck, I might be able to observe her and the cubs for a longer period of time now that I knew she was there, provided, of course, that I traveled slowly and as silently as possible.

Searching for, and finding, a wild animal is my favorite sport, an avocation, almost. It requires extreme patience, a detailed knowledge of the quarry, and the kind of commitment that sharpens the faculties to a point where it is difficult to determine whether one is being motivated by conscious concentration or by instinctive and primitive senses. It is certainly a fact that ears, nose, and eyes become more efficient and that the feet seem to develop sight of their own, or at least a sense of touch, that allows one to avoid stepping on dry twigs or stumbling against obstacles, while the eyes are searching the ground some distance ahead. Walking blind, I call it. I first became aware of it many years earlier when, on night patrol in the war, I discovered that my mind could sense the contours of the land and somehow guide my feet; I can't explain the phenomenon, but it is similar to the instinct that helps blind persons to "see" their way.

One does not travel fast when engaged in such a quest, but one becomes so fascinated by it and by the many things that one notices that speed is unimportant; time has no meaning. It doesn't even matter a great deal if one fails to sight the quarry, for it is the *doing* that counts, though that is not to say that the successful completion of a good stalk does not bring a great deal of personal satisfaction; indeed it does! There are few experiences that give me a greater thrill than the first sighting of an animal that I have been stalking for hours, especially if I can come upon it before it has detected me and can thus observe it at leisure. That's the kind of hunting that I am partial to, bloodless and rewarding. Not only do I learn something about the quarry by observing it, but I make other discoveries at the same time, little nuggets of knowledge picked up on the way that are coincidental to the objective but that provide new insight.

I failed to find the female bear and her cubs that afternoon and I was quite late getting home because of the stalk, but if I was to be scolded by my wife for once again missing the supper hour, I was to be rewarded by a delightful experience.

By the time that I reached our own maples it was night, and an overcast that had moved in during late afternoon was obscuring the stars. In total darkness I came out of the hardwoods and began to cross the clearing, guided now by the lights in our home. About halfway across the opening I detoured slightly to avoid the rutted tractor trail that led to the evaporator house, moving closer to a line of elms. As I neared the trees, something tapped me lightly on the head, just once. There was no noise that I was able to detect, nothing in sight, merely that feather-soft touch, the pressure of which was immediately eased; when I put my hand up to feel my hat, there was nothing on it. It was a decidedly eerie experience, its very unaccountability giving it almost supernatural connotations, particularly in view of the fact that I was walking in the open, where there was no chance of brushing against a low branch or of a twig falling on top of me.

I was about ready to believe that I must have imagined the incident when something definitely touched me again, this time on my left shoulder, but the contact was not on this occasion immediately broken. Moving slowly, I raised my right hand and allowed it to descend gently on the place of impact, but on feeling a warm, extremely smooth object that vibrated and uttered a small squeak, I was unable to control the reflexes that jerked my hand away as quickly as if it had come into contact with a red-hot surface.

My involuntary reaction produced a second squeak from the invisible night sprite accompanied by movement so swift that I would not have believed it possible had I not experienced it on my own body. But the little cry, and the lightninglike scurry from shoulder to a place immediately under my chin, furnished me with the identity of my unusual passenger: a flying squirrel. As if to confirm my recognition, the little animal's downy tail rose up and brushed across my lips and nose.

I don't startle easily under any circumstances, but I must confess that the squirrel's unseen, unheard, and sudden choice of landing site did cause the hairs on the nape of my neck to erect themselves, if only momentarily. Such an unorthodox visit had not occurred before, a fact that led me to guess that the flying squirrel perched now on my sternum was George, the most inquisitive and companionable of the gang. Reaching into my parka pocket for a peanut, I spoke to my visitor, and he responded by chirping at me again, a soft, musical little note, and by hopscotching down my parka, aiming toward my seed pocket. Even as my hand started to emerge with the peanut, George's soft, cold nose pressed itself against my skin. I didn't complete the move, merely cupping the nut loosely in a partly closed fist; a second later George's little head pushed in and his mouth secured the nut. Hopscotching upward again until he was on top of my head, he launched himself into the dark and I heard the soft "plump" of his landing on the bole of a nearby elm.

I count that as one of the most remarkable encounters with a wild animal that I have ever experienced. It was a thrilling, spiritual meeting on a dark night that convinced me beyond all doubt that man, provided he doesn't harbor violence in his soul, can reach across the centuries of civilization and commune with the other living entities with whom he shares the world. George came to me unbidden because he knew that I was a friend and a source of food that would be freely offered, even as he shared with his own kind. And having crossed the barrier that separates man from animal, George continued to trust me fully from then on, making it a habit to alight on my person and to accept a peanut, or even merely to sit and clean himself perched on shoulder or head, responding eagerly to my scratching fingers and arching his neck when I stroked it gently with an index finger.

Flying squirrels are found practically all over North America and even as far south as parts of Mexico and Guatemala; yet few people realize that they are present in their neighborhood. This is due to the animal's almost totally nocturnal habits and to the remarkably silent way that it moves about at speeds that can baffle the eye. Measuring between nine and fourteen inches from tip to tip and weighing from one and a half to a maximum of five and three-quarter ounces, the little squirrels subsist on a varied diet, taking seeds, fruits and nuts, and even meat when this is available. Social in habit, they think nothing of feeding as a group, something that would cause a riot if attempted by the red squirrels.

Their genus-specific name, *Glaucomys volans*, means "silver gray mouse that flies" and is as misleading as their English name, actually more so! Far from being silver gray, these squirrels are light brown to fawn on back and sides and very light fawn, almost white, on the underparts. And they don't fly! They glide, being able to do so with the aid of twin folds of loose skin that extend from ankle to wrist. When the squirrel jumps from the top of a tree, it spreads all four legs wide and the loose skin taughtens and catches the wind, converting its body into a perfect glider. The hairs on the little animal's tail are flattened and grow rather like the veins of a feather, allowing *Glaucomys* to use this appendage as a rudder as well as a balancing organ. By varying the angle of its legs, it can alter the slack in one or both of its "wings" and in this way it can obtain additional control of its course, speed, and angle of descent. Of course, when it is on the ground, or low in a tree, the squirrel cannot flap away like a bird but must climb again to the topmost branches if it wants to make a big glide that may cover two hundred feet or more. For shorter "hops" it springs into space, opens its flaps, and lands almost at once, gaining as much momentum from the kickoff as from the "wings."

George's trust encouraged his kin to accept me more fully, even if none of the others ever dared to land on me. Yet they would follow my little friend and would hop down the trunk of a tree to accept peanuts from my fingers. Because of this, I resumed my study of the species and took to spending quite a lot of time outdoors at night, especially during the times of full moon. My observations continued well into July. It was while thus engaged that I met Snuffles again.

I was sitting in a garden chair behind the house holding a dish of peanuts and sunflower seeds in my lap. George came at intervals to help himself, and his relatives chirped and squeaked from the vantage of the elm tree under which I was stationed, each little fawn animal scurrying down the trunk, there to hang, looking at me until I offered a peanut. The time was about 11 P.M., the date, July 14, and the moon was full in a clear sky. George was sitting inside the bowl deftly shelling and eating sunflowers when he suddenly sprang into the tree and disappeared, his hurried departure warning me that a predator was somewhere in our vicinity. As I listened, Tundra, chained to the porch, lunged at his fetters and yiped excitedly; he, too, had been alerted, but my own poor ears were still unable to pick up any noise. I remained in the chair, unmoving.

Some moments later I at last heard a sound, a soft snuffle that was almost as familiar as my own breathing. Directing my gaze toward the place from which the noise had come, I noted the dark, amorphous shape of Snuffles. He was moving toward me, but slowly, as though not quite sure of himself. Not wishing to startle him, I spoke his name quietly and stood up, stepping away from the tree's shadow and into the open.

Snuffles stopped, snuffled a couple of times, and snorted softly. Tundra barked his high and repetitive cry of anger and lunged at his chain again, but neither the rattle nor the bark elicited response from the bear. He ambled toward me, snuffling continually and causing me to realize that I was holding the dish of peanuts and seeds, the scent of which undoubtedly had reached his keen nose.

I didn't want to encourage him but at the same time I was pleased that he had come to visit. It would be ungracious, I felt, to send him away without offering him a little something. Calling him again, I moved toward him, stopped when we were some five yards apart, and emptied the contents of the bowl on the grass. Snuffles padded up, lowered his head, and started eating, now and then raising his face and looking at me.

The sound of the windowpane sliding open startled him, but before Snuffles had time to wheel around and run, Joan's voice reached into the night.

"Is that my Snuffles?"

Turning, I noted that my wife's head and shoulders were sticking out of the window, signifying that she was practically lying on top of the counter. In her right hand she held several pieces of bread, evidently hastily picked up when she became aware of the bear's presence, alerted no doubt by Tundra's noise.

Snuffles marched right up to the window, dignifying me with but a quick glance as he passed by, then reared to his full and impressive height when only two feet from the feeder. His nearness caused Joan to pop herself back inside, but she extended the bread and Snuffles leaned forward to take one slice in his mouth. Pivoting like an obese dancer, he dropped to all fours, marched to a spot some six feet distant from the house wall, flopped down on the grass, and began to eat the gift daintily, taking his time. I called to Joan, suggesting that she might like to make him a honey sandwich or two.

The outcome of that visit was that Snuffles accepted four honey sandwiches, half a package of marshmallows, and about a pound of peanuts, comporting himself like a perfect gentleman and marching back into the wilderness at the end of an hour, his behavior, and my wife's, reminding me once again of the unpredictability of wild animals—to say nothing of Joan!

When she had given Snuffles his first honey sandwich, Joan came out of the house and walked around to the back, carrying the other good things. One by one she fed the bear his "gifts," squatting in front of him without showing anything but love. After he had eaten the lot and remained lying down slurping the last vestiges from his lips, Joan reached out a hand and scratched his head, uttering nauseating endearments while doing so. And that damned great, ugly beast rolled over on his back so that she could scratch his chest and belly!

I protested, of course, cautioning Joan against such risky familiarity, but I was ignored. In due time my wife arose and came to stand beside me and soon thereafter Snuffles lumbered to his feet, snuffled once, and turned his back on us, ambling away to be quickly swallowed by the gloom of the forest.

That visit established a pattern. At irregular intervals, usually several weeks apart, Snuffles would come to visit, but always at night. Habitually he approached the house from the rear, stood up in front of the feeder, and ate whatever remained on it. If we were in bed, he would tell us he was there by gnawing on the wooden shelf, the rasping of his great teeth rousing Tundra to fever pitch, the dog's latration awakening us. We then would go downstairs, open the window, and speak to the bear, while Joan made up a quota of honey sandwiches. After this the marshmallows were offered and accepted. Last, but an important part of the ceremony, Joan would put a mound of peanuts on the shelf and scratch the bear's head while he ate them.

In time, Snuffles was to wander away into the deep forest, never to return; this was as it should be. The memory of him and the friendship he had given us were our rewards for rescuing him. We were not sad when he found his rightful place in the wildwoods and we counted the time we had spent caring for him as richly employed.

Meanwhile, The Zoo That Never Was continued to be. Animals came, stayed awhile, left. Others died in our sanctuary, but they did so in peace. And there were those that remained, living wild and free on our acres and giving us their trust and friendship.

Discussing our experiences one winter's night as we sat before the fire attended by Tundra, we realized that the companionship and trust that had developed between us and the animals that we had befriended had allowed us to experience a oneness with nature that was inestimably fulfilling. What began as an interesting, if somewhat onerous, task, undertaken rather casually, had ripened into thoughtful commitment that taught us much about ourselves and about the value and sanctity of all life.

We were silent for a time, holding hands, communing with each other and with our great dog, who sprawled between us with a smile wreathing his countenance, his ivory teeth gleaming. Staring into the dancing flames of the fire, a thought that was uttered more than two thousand years earlier by a Roman whose name I couldn't recall surfaced unbidden in my mind. It summed up our feelings so aptly that I quoted it aloud: "Vitaque mancipio nulli datur, omnibus usu."

I was hardly aware of speaking until Joan asked me to translate.

"I said, my love, that life is given to no one absolutely, but that everyone has a right to use it."